DATE DUE

IY 2 0 '12		
FE 1 2 '14		
SE 2 4 '15		

Understanding Vineyard Soils

Understanding Vineyard Soils

Robert E. White

OXFORD
UNIVERSITY PRESS
2009

OXFORD
UNIVERSITY PRESS

Oxford University Press, Inc., publishes works that further
Oxford University's objective of excellence
in research, scholarship, and education.

Oxford   New York
Auckland   Cape Town   Dar es Salaam   Hong Kong   Karachi
Kuala Lumpur   Madrid   Melbourne   Mexico City   Nairobi
New Delhi   Shanghai   Taipei   Toronto

With offices in
Argentina   Austria   Brazil   Chile   Czech Republic   France   Greece
Guatemala   Hungary   Italy   Japan   Poland   Portugal   Singapore
South Korea   Switzerland   Thailand   Turkey   Ukraine   Vietnam

Published by Oxford University Press, Inc.
198 Madison Avenue, New York, New York 10016

www.oup.com

Oxford is a registered trademark of Oxford University Press.

Library of Congress Cataloging-in-Publication Data
White, R. E. (Robert E.)
Understanding vineyard soils / Robert E. White.
    p. cm.
Includes bibliographical references and index.
ISBN 978-0-19-531125-9
1. Grapes—Soils. 2. Viticulture. I. Title.
S597.G68W 45 2009
634.8'8—dc22   2008029797

9 8 7 6 5 4
Printed in the United States of America
on acid-free paper

Preface

More than any other product of the land, the flavors and aromas of wine are linked to the soil, a linkage that is central to the concept of *terroir*. This book is not about *terroir*—that is a subject for wine writers to expound on, and for wine makers and vignerons to wax lyrical about on Web sites and wine labels. What this book is about is the basic properties of soil: how they vary from place to place and why this variation dictates how growers should manage the soil in their vineyards to achieve specific objectives of yield, grape quality, wine style, and personality.

I have tried to keep the book simple, but not so simple that an inquisitive reader who wants to understand why certain things happen cannot delve more deeply by exploring the boxes in the text and pursuing references.

In gathering material for this book I have been privileged to have assistance from many colleagues and friends. I would particularly like to thank Rob Bramley, Tony Proffitt, Kees van Leeuwen, Judy Tisdall, and Kevin Bell, who read one or more chapters. Mark Walpole and Brad Johnston kindly commented on the outline proposal for the book. My wife, Annette, gave constant support on our many enjoyable travels to wine regions and during the exacting task of writing. I am also most grateful to those people, individually acknowledged, who provided me with illustrations and other data.

Nevertheless, despite their best efforts, the final responsibility for the content and interpretations rests with me.

Robert E. White
Melbourne, April 2008

Contents

Understanding Vineyard Soils

1

Why Soil Is Important in Viticulture

The Pétrus secret begins with soil.

▓ From Clive Coates (2005), *The Great Wines of France: France's Top Domaines and Their Wines*

A Global Snapshot

Throughout the world, grapevines grow in all kinds of soil. The factors controlling soil formation are universally present, but their individual influence and subtle interactions vary greatly from place to place, giving rise to soil variability and management challenges.

Consider, for example, a winegrower on the Willakenzie soil in the Willamette Valley, Oregon, USA, who complains that water tends to perch on the impermeable subsoil during wet winters only to dry out too quickly in the topsoil during the summer. Such a condition also occurs in some of the duplex soils in southeastern Australia and other cool-climate regions with similar soils and marked seasonal contrasts. This combination of soil profile form and climate creates problems for a vineyard, such as poor drainage (figure 1.1). On the other hand, vines grown on the deep red loams on basaltic rock material in the Willamette Valley tend to be too vigorous. Although well drained, these soils are naturally fertile and supply abundant water, a condition akin to that of the deep red loams in the upper King Valley and parts of the Yarra Valley, Victoria, Australia, which have a climate similar to the Willamette Valley. Again, the combination of soil and climate poses specific management problems. Figure 1.2 shows a comparison of these red loam soil profiles; box 1.1 describes what is meant by a soil profile and highlights some of the main variations seen therein.

In contrast to these deep fertile soils, vines struggle on soils where soil development is limited by the parent rock. For example, soils formed on the metamorphic

Figure 1.1 A soil with an impermeable subsoil in Virginia, USA. The vines in this vineyard died from waterlogging. (Photo courtesy of Brad Johnston, Warrenton, Virginia, USA)

rock schist support vineyards in Central Otago, New Zealand, and the Coteaux du Languedoc, France (figure 1.3). Here the soils are shallow, very stony, and excessively drained. If not irrigated, the vine roots must penetrate deeply into the fissures of the thinly bedded rock in search of water. The soil constraints can be overcome to some extent by deep ripping at vineyard establishment (see chapter 2).

Soils formed on limestone provide another example where root development can be inhibited. A classic example is the calcareous brown soil formed on the hard limestone of the Côte d'Or in Burgundy, France. In its natural state, this soil has an organic-rich surface layer (figure 1.4A), but cultivation throughout several centuries has all but removed the organic horizon and has left a shallow mineral soil (figure 1.4B). However, some roots do penetrate down fissures in the limestone, which slowly releases water to the vine (figure 1.5). Other examples occur in St. Emilion, France, the Paso Robles region of California, USA, and the upper slopes of McLaren Vale, South Australia. In the latter, deep ripping is necessary to break up the limestone that occurs at a shallow depth (figure 1.6).

Many vineyards are found on soils that are deep and well drained, but not very fertile; thus, they require a range of nutrient inputs. Examples are the gravelly

(A)

(B)

Figure 1.2 (A) A deep red loam on colluvial basalt in a vineyard of the Willamette Valley, Oregon, USA. (B) A deep fertile red loam on basalt under vineyards in the upper King Valley, Victoria, Australia.

Box 1.1 The Soil Profile

A vertical exposure of soil in a pit or road cutting is called a soil profile. Often there are obvious changes in color and composition from the surface (enriched with organic matter from plant litter) to the subsoil and parent material below, as can be seen in the soil profile in figure B1.1.1. When these changes in color and composition are the result of soil formation in different rock materials, the profile is called layered. Often, a lower layer represents a buried or "fossil" soil (figure B1.1.2).

    When profile changes are the result of different chemical and biological processes occurring in the same parent material, the layers are called horizons and are labeled A, B, and C, from top to bottom. Horizons may be subdivided. For example, in figure B1.1.1, the upper zone of the A horizon, which is rich in dark-brown organic matter, is labeled A1 to note the distinction from the paler zone immediately below, which is labeled A2. The A and B horizons comprise the soil proper, and the C horizon consists of weathered parent material. An O horizon is a superficial organic horizon composed of partially decomposed litter.

Figure B1.1.1 Soil profile in the Pyrenees wine region, Victoria, Australia, showing A and B horizons. The scale is 15 cm.

(continued)

Box 1.1 *(continued)*

Figure B1.1.2 Soil layers comprising a buried or fossil soil in the Riverina wine region, New South Wales, Australia. (White, 2006; reprinted with permission of Wiley-Blackwell Publishing Ltd.)

Broadly, soil profiles may be divided into the following categories:

- Uniform: little change in texture with depth (e.g., figure B1.1.3)
- Gradational: a gradual increase in clay content with depth (e.g., figure 1.2B)
- Duplex or texture contrast: a change from a "light" textured A horizon (sandy loam to sandy clay loam) to a "heavy" textured B horizon (clay loam or clay), usually with an abrupt boundary between the two (e.g., figure B1.1.1)

The A horizon (topsoil) is easy to distinguish from the B horizon (subsoil) in a duplex profile, but this separation is less obvious in a gradational profile, and even less so in a uniform profile. In the latter cases, a distinction between A and B horizons can be made on the basis of organic matter content, which is

(continued)

Box 1.1 *(continued)*

Figure B1.1.3 A deep, uniform soil profile formed on granite in the Calquenas wine region, Chile. (White, 2003)

greater in the topsoil and usually decreases to an insignificant amount below 20 cm or so.

soils of the Bourdon Ranch area in Lodi, California; the Gimblett Gravels of Hawkes Bay, New Zealand; and the Wairau Plain of Marlborough, New Zealand (figure 1.7). These soils are formed on transported parent materials, deposited to considerable depth by rivers throughout thousands of years. The stones, and even boulders, have been worn smooth and rounded by the action of water. The out-wash material from glaciers during past ice ages is similar (box 1.2), except that it is often more mixed in size and rock type, and is not as well rounded as the river deposits. Such deposits occur in the Rhone Valley and Médoc, France, and in several winegrowing valleys in Chile (figure 1.8).

These few examples illustrate the diversity of the influence of parent material, climate, topography, organisms, and time (summarized in table 1.1) in determining the direction of soil formation. The resulting variability—expressed through soil properties such as depth, profile form, structure, drainage, water storage

(A)

(B)

Figure 1.3 (A) A vineyard on a shallow soil on schist in the Central Otago wine region, New Zealand. Note the schist outcrops in the background. (B) Old bush vines on a schistose soil in Coteaux du Languedoc, France. (White, 2003)

(A)

(B)

Figure 1.4 (A) An organic-rich shallow soil on limestone in the Côte d'Or, France. (White, 2003)
(B) A soil similar to that in A, but with little organic matter remaining after centuries of cultivation.

Figure 1.5 Fractured limestone rock underlying Grand Cru vineyards near Gevry-Chambertin in the Côte d'Or, France. The scale is 15 cm. (White, 2003)

Figure 1.6 An organic-rich soil developed on limestone in McLaren Vale, South Australia. The scale is 20 cm. (White, 2003)

Figure 1.7 A soil developed on a gravel and boulder alluvium in the Marlborough wine region, New Zealand. The scale is 10 cm.

capacity, presence or absence of salts, biology, and fertility—must be taken into account when managing soil in a vineyard. Success in the production of an optimum yield of quality grapes, with minimum impact on the environment, will depend on how well vineyard managers understand soil variability and are able to work with this variability, or are able to modify it to their own ends.

How Soil Forms

Even where vines appear to be growing in rock, we know that, initially, a soil—no matter how meager—was present, because when a rock is first exposed to weathering, a soil begins to form. Figure 1.9 shows a very early stage of soil formation on granite.

Soil is distinguished from pieces of rock by the presence of living organisms and organic matter. An assortment of inorganic minerals inherited from the parent rock forms the structural framework that provides living space for a diversity of organisms. Life in the soil is a struggle, but the cycle of growth, death, and

Box 1.2 The Geological Time Scale

Compared with the 6000 years or so of recorded human history, Earth's age of 4.6 billion years is unimaginably long. Nevertheless, events that occurred millions of years ago have relevance today because of the unending cycle of rock formation, weathering, erosion, and deposition that have laid the foundations of, and helped to shape, the current surface of Earth. As the science of rocks (geology) developed, the history of Earth's rocks was divided into a timescale of eras, periods, and epochs. Much information is available for rocks that date from the appearance of the first multicellular life forms some 650 million years ago through many subsequent epochs to the present. Periods within eras are usually associated with sequences of sedimentary rocks that were deposited in the area now known as Europe. Although examples of these rocks are found elsewhere, the European time divisions have generally been accepted worldwide, except in cases such as the Ediacarian period of the Pre-Cambrian era, whose defining fossils were first identified in the Flinders ranges of South Australia. Table B1.2.1 shows a simplified geological timescale from the Cambrian period, when life forms exploded in diversity and number, to the present.

Table B1.2.1 The Geological Timescale

Era	Period	Epoch	Start time (million years ago)
Cainozoic ("young life")	Quaternary	Recent	0.011
		Pleistocene (the ice ages)	2
	Tertiary (the Age of Mammals)	Pliocene	5
		Miocene	23
		Oligocene	36
		Eocene	53
		Paleocene	65
Mesozoic ("middle life" and the Age of Reptiles)	Cretaceous		145
	Jurassic		205
	Triassic		250
Paleozoic ("old life")	Permian		290
	Carboniferous		360
	Devonian		405
	Silurian		436
	Ordovician		510
	Cambrian		560
Pre-Cambrian			4600

Figure 1.8 Soil developed on a glacial outwash deposit overlying older weathering granite in the Calquenas wine region, Chile. The scale is 15 cm. (White, 2003)

decay is essential for healthy soil functioning in cultivated fields, forests, grass-lands, and vineyards. Organisms ranging from bacteria (invisible to the human eye), to insects, earthworms, and burrowing animals feed on the residue of plants and other organisms, and themselves become food for subsequent generations (see chapter 5). An answer to the question "Why does the course of soil formation diverge so much in the landscape?" lies in our understanding of how the chemical, physical, and biological processes of profile development vary under the multifac-eted influence of soil-forming factors. Table 1.2 identifies some of the key profile-forming processes that occur naturally.

Soil Variability and the Concept of *Terroir*

For centuries, wine makers have been aware of the interaction between soil var-iation, the performance of particular cultivars or varieties, and wine character. The best recorded examples go back to the Middle Ages (from the eighth century AD) in Burgundy, and the Rheingau, Germany, when monasteries possessed large

Table 1.1 Summary of Soil-Forming Factors and Their Effects

Factors	Components	Characteristics and active agents	Features of soils formed
Parent material	*Parent material in place*		
	Igneous rocks—derived from volcanic activity	Acidic rocks (e.g., granite, rhyolite), rich in quartz and pale-colored Ca^a, K^a, and Na^a feldspars	Soils on granite have a high proportion of quartz grains, often coarse; well drained if deeply weathered; low fertility
		Basic rocks (e.g., basalt, dolerite), low in quartz and rich in dark-colored ferromagnesian minerals	Soils on basalt often high in clay and Fe and Al oxides; can be poorly drained unless old and well weathered; generally fertile
	Sedimentary rocks—weathered rock material deposited under water or by wind and subsequently compressed	Varying size of rock particles forming mudstones (small), sandstones, shales, and conglomerates (large); fine-grained chalk and limestone	Soils on mudstone often poorly drained; soils on sandstone and conglomerates of low fertility; soils on limestone vary, based on content of impurities
	Metamorphic rocks—igneous and sedimentary rocks subjected to heat and pressure that has changed the original rock structure and minerals	Granite metamorphosed to gneiss, sandstone to quartzite, shale to schist, limestone to marble	Variable weathering and chemical properties, depending on the original rock that was metamorphosed
	Transported parent material		
	Weathered rock materials moved from their original place	Water, gravity, ice, and wind deposits; when compressed, can form sedimentary rock	Alluvial (water) and colluvial soils (gravity), loess (wind), usually well drained and fertile
Climate	Macroclimate—the regional climate applicable to a subregion or site, such as a valley or slope	Modified by latitude and elevation; temperature affects the rate of organic matter decomposition and other biochemical reactions; moisture availability affects weathering, leaching, and clay movement down the soil profile	Current and past climates are reflected in present-day and buried ("fossil") soils through differences in depth, color, organic matter, structure, translocation of clay, organic compounds, salts and oxides, and ultimately the profile form (see box 1.1)
	Mesoclimate—applicable to a subregion or site, such as a valley or slope		
	Microclimate—very local, as within the canopy of a vineyard		

(continued)

Table 1.1 (*continued*)

Factors	Components	Characteristics and active agents	Features of soils formed
Organisms	Plants Animals Humans	Plant species differ in the chemical properties of their leaf litter; roots and earthworms form channels; microorganisms decompose, and humans add, organic residue	Under conifers and heath, Fe and Al organic complexes translocated to form podzols and podzolic soils; humified organic matter profiles form under deciduous trees and grassland; humans form "plaggen" soils
Topography	Slope	Water and gravity remove material down steep slopes; solifluction occurs in thawed over frozen soil	Soils at the top of slopes are shallow with good drainage; depth increases downslope but drainage usually is poorer; Fe colors change from orange-red through yellow to blue-gray downslope
	Aspect	Slopes facing the equator are warmest	
Time	Tens to hundreds of years (recent volcanic eruptions, human activity) to tens of thousands of years (glacial and interglacial periods of the Pleistocene [box 1.2]) to more than one hundred thousand years (stable landscapes)	Modifies the action of the other factors	Soils range from shallow with little profile development to deep with well-developed profiles as time passes; younger soils are generally more fertile

[a] Chemical symbols for the elements are explained in chapter 3.

Figure 1.9 A thin mantle of soil forming on granite under early-colonizing plants. The scale is 10 cm.

tracts of land on which vines were grown and wine was made. Particularly notable were monasteries at Cluny in the Mâconnais, France, Clos de Vougeot at Cîteaux in the Côte d'Or, and Kloster at Eberbach in the Rheingau, where successive generations of monks identified, by trial and error, the special character of wine made from individual parcels of land (*goût de terroir*). Cultivars were winnowed out until the first choice was Pinot Noir in the Côte d'Or and Riesling in the Rheingau. Gamay was found unsuitable on the soils of the Côte d'Or and was "banished" by order of Duke Philippe of Burgundy during the late 14th century, but subsequently found its natural home on the schist and granite soils of Beaujolais to the south. Subsequently, Chardonnay was planted on the hard limestone soils of the southern Côte d'Or, extending into the Chalonnais, and farther south to parts of the Mâconnais, where the beautifully structured calcareous brown soils around the Roche de Solutré and Roche de Vergisson produce excellent white wines (figure 1.11).

Having identified the best combinations of cultivar and soil, the wine makers strove to maintain consistency in quality and character for individual labels by blending wine from small blocks in different proportions, depending on the seasonal conditions. Thus were laid the foundations of the concept of *terroir*, which expresses in one word the complex interaction of soil, local climate, cultivar, and the wine maker's skill in determining the character or individual "personality" of a wine. After the French Revolution of 1789, Napoleon ordered the large monastic landholdings to be broken up, with the result that vineyards were fragmented into 0.5- to 1-ha blocks (called *climats*) and were acquired by many owners. However, much of the unofficial classification of wine from the *climats* in Burgundy was preserved and incorporated in 1861 into an official Appellation d'Origine Contrôllée

Table 1.2 **Summary of the Main Soil Profile-Forming Processes**

Process	Chemical reactions and their outcomes
Solution and hydrolysis	Water is active in mineral weathering; soluble salts such as sodium chloride (NaCl) and even less soluble salts such as gypsum ($CaSO_4 \cdot 2H_2O$) dissolve. Water hydrolyzes silicate minerals by splitting into H^+ and OH^- ions, with the H^+ ions attacking the minerals, and clay minerals are formed (see chapter 3).
Oxidation and reduction	Weathering of Fe-containing minerals releases ferric (Fe^{3+})[a] and ferrous (Fe^{2+})[a] ions, which are in a reversible equilibrium depending on O_2 availability. High O_2 concentrations (aerobic conditions) favor Fe^{3+} compounds, particularly the oxides ($Fe_2O_3 \cdot nH_2O$), which are insoluble and orange-red ("rusty"). Low O_2 concentrations (anaerobic conditions) favor Fe^{2+} compounds, which are more soluble than Fe^{3+} compounds and give a blue-gray color to the soil. Soils that are waterlogged most of the time are predominantly blue-gray, whereas those that are periodically waterlogged have rusty Fe "mottles."
Humification	As organic residues are decomposed by microorganisms, the material becomes humified. The black, amorphous end product is called humus (see chapters 3 and 5). On neutral or calcareous parent materials, the humus formed is called mull and is well mixed into the mineral soil by earthworms and other burrowing animals (fig. 1.10A). On acidic parent materials, especially under conifers and heath, mor humus several centimeters thick forms with a sharp boundary separating it from the mineral soil below.
Complex formation	Reducing compounds such as polyphenols are leached from leaf litter and mor humus, and form complexes with Fe and Al. These complexes are sufficiently stable to be leached down the soil profile until they break down at depth. Bands of organic matter and orange-red Fe oxides build up in the subsoil (fig. 1.10B). Typically, a bleached horizon develops in the zone from which Fe and Al have been leached.
Leaching	Ions dissolved from soluble salts such as NaCl are readily leached. Cations such as Ca^{2+}, Mg^{2+}, and K^+ do not leach as readily as Na^+. Anions such as SO_4^{2-}, HCO_3^-, and NO_3^- also leach, and there is the possibility of calcium carbonate ($CaCO_3$) and gypsum precipitating deeper in the profile, depending on the excess of rainfall over evaporation. Iron and Al oxides are very insoluble, and the metal ions do not leach unless they are complexed or Fe is reduced.
Translocation	Over a long period of time, clay-size particles are moved downward by water and deposited deeper in the soil profile, which gives rise to a B horizon. Typically, this is how a duplex soil forms (see box 1.1). Clay is more likely to be translocated when it carries Na^+ ions, whereas it is much more stable when Ca^{2+} ions predominate (see chapter 3).

[a] Ions and their charges are explained in chapter 3.

(A)

(B)

Figure 1.10 (A) The dark
mull humus layer of a soil
on limestone in the Paso
Robles region, California,
USA. (B) A deep sandy
profile showing a bleached
horizon above translocated
organic matter and an
orange-red, iron oxide
subsoil. The pale soil on
top is spoil from the pit.

Figure 1.11 A beautifully structured brown soil near the Roche
de Solutré in southern Mâconnais,
France. The scale is 10 cm.

(AOC) system, which still exists today. Now there are several hundred appella-
tions in the greater Burgundy region.

The French AOC system expresses the concept of *terroir* through an identification of viticultural areas, based on the distinctive character of wines traditionally produced in those areas. The AOC system prescribes the cultivars, viticultural
methods, maximum yield, and wine alcohol content that are acceptable in a particular appellation, and the wines produced are allocated to four categories from
the top down: Appellation Contrôllée, Vin Délimité de Qualité Supérieure, Vin
de Pays, and Vin de Table.

The strict AOC regulations make it difficult for producers to experiment with
new cultivars and production systems, and the regulations and boundaries are slow
to change in response to new developments in grape growing and wine making.
Consequently, some enterprising wine makers have abandoned the AOC system
and the similarly restrictive Denominazione di Origine Controllata e Garantita system in Italy, to produce successful wines such as the Super Tuscans in the Chianti
region of Tuscany, and the *vin de cépages* of Languedoc-Roussillon, France.

Despite the reputation of many of the top wines in the AOC system, there is
much wine of ordinary quality marketed in the lower categories. Overproduction
in France and fierce international competition in the middle and lower price

ranges have led to a relaxation of AOC regulations in some French wine regions. Also, New World countries have adopted regional classification schemes, such as the American Viticultural Areas and the Geographical Indications in Australia, which go some way toward acknowledging the concept of *terroir*, or "sense of place," as it is referred to in Australia.

Outside the well-established wine regions, the practical assessment of a new site's *terroir* is hampered by the fact that the causal relationships between particular soil properties and the *terroir* are not known. A possible exception to this generalization is the relationship between the rate of soil water supply and the performance of dryland vines, which has been extensively studied in the Médoc and St. Emilion regions of Bordeaux (see chapter 6). Thus, in the absence of such knowledge, a pragmatic approach is to think of *terroir* as an empirical classification specifically for vineyard soils, which has been honed by the experience of practicing vignerons over a long time.

Soil Classification and *Terroir*

Classification involves the "packaging" of soil variability so that soils with similar property values such as profile form, color, and texture are grouped in one class, and are separated from soils with dissimilar property values. To assess the full range of soil variation in an area of interest, a survey is conducted. The variation is divided into classes, which collectively constitute a classification. Soil scientists have justified this kind of traditional classification by saying it is valuable for communicating soil information nationally and internationally. However, experience shows that this information has not been widely used by farmers, growers, and land managers, in part because the detailed information about a particular location is lost in the creation of the general-purpose classes, and in part because scientists have produced several classifications that have resulted in a confusing suite of soil names. In Australia, for example, four national soil classifications have been used, culminating in the Australian Soil Classification (Isbell, 1996), which was revised in 2002 (Isbell, 2002). When it is necessary to use a general soil name in this book, I use one from the older Great Soil Group Classification, which retains some currency in the viticulture industry.

Recognition of *terroir* in Old World vineyards, especially in Burgundy and other historically important regions of France, relies on an accumulated knowledge of soil properties relevant to vine performance and fruit quality on a scale of meters rather than kilometers. In Australia, where the wine industry is much younger than in France, the need to provide relevant soil data for grapevine culture has led to the creation of a special-purpose classification, which as far as possible uses nontechnical terms to describe soil properties observable in the vineyard (Maschmedt et al., 2002). However, the next step in soil classification is to provide these data at an appropriate scale, which is where modern techniques of sensing a soil property at close range or remotely come into their own.

Soil Survey and Mapping

Traditionally, a soil survey involves a preliminary examination of samples collected by "augering" or coring. Some examples of soil augers and corers are shown in figure 1.12. After the boundaries separating classes have been sketched in, rectangular pits are dug to expose the soil profile thought to be representative of each class. Figure 1.10B is an example of one such profile pit. The profile pit allows the visual characteristics of each horizon to be described accurately; these include sharpness of boundaries, depth, color and mottling, presence of organic matter, stones, structure, and presence of carbonate or other salt deposits, as well as the "feel" or texture of the soil (described in detail in chapter 2). Samples are also collected from each horizon for laboratory analysis. The "standard" practice for a vineyard soil survey is to locate pits on a 75 × 75-m grid.

The output of such a survey is a map that shows the spatial distribution of soil classes. If the soil distribution is relatively simple, or the map scale is large, individual classes can be displayed. However, if the distribution is complex (as it often is

Figure 1.12 Examples of augers and corers for soil sampling. (White, 2006; reprinted with permission of Wiley-Blackwell Publishing Ltd.)

Order
■ Vertosol
▨ Tenosol
▨ Rudosol
▨ Kurosol
□ Kandosol
■ Hydrosol
■ Ferrosol
■ Dermosol
▨ Chromosol
■ Anthroposol

Figure 1.13 A digital map of topography and soils in the Pokolbin district of the Hunter Valley wine region, New South Wales, Australia. (Courtesy of the Australian Centre for Precision Agriculture, Sydney, Australia)

with alluvium-derived soils) or the map scale is small, two or three classes may be grouped into a mapping unit. From such a conventional survey, which is usually linked to a general-purpose soil classification, the key properties of a soil typical of each class are identified. This typical soil is called a soil type, and the properties of each type are supplied in a legend that accompanies the soil map. Because such soil maps are rarely available at a scale larger than 1:5000[a] (and more commonly at 1:50,000), much detail about soil variation over distances of a few meters is lost.

However, it is now possible to conduct a survey by measuring soil properties in a dense spatial array in which the measurement points are accurately located by a global positioning system (GPS). The GPS relies on fixing a point on the surface with reference to orbiting satellites, and the best system in general use—the differentially corrected GPS or dGPS—is accurate to about ±50 cm in Australia. When a dGPS is combined with a real-time sensing instrument, data can be rapidly collected in, say, a 2 × 2-m grid, which provides 2500 values of a soil property per hectare. An even more sophisticated system is the real-time kinematic (RTK) GPS, which is accurate to 2–3 cm in the horizontal and vertical planes. When used with a height reference level and a geographic information system (GIS; described later), the RTK GPS enables a digital elevation model (DEM) of the site to be constructed and displayed in a three-dimensional map. The soil distribution, based on measured properties and a model of soil formation, can then be draped over the DEM and displayed as shown in figure 1.13.

[a] This scale means that 1 cm on the map represents 50 m in the field.

A number of sensing instruments that operate close to the soil (proximal) or remotely (from the air) are now available or being developed—for example, for sensing pH, soil water content, electrical conductivity (*EC*), penetrometer resistance, texture change, and visual properties such as color (including mottling) and cracking. More comprehensive information can often be obtained with a multisensor, such as a cone penetrometer to measure soil resistance (related to texture and bulk density) combined with an imaging penetrometer that registers soil color. As the instrument is pushed into the soil, physical properties and color are recorded for subsequent analysis, which allows the presence of sand and clay layers and horizon boundaries to be detected (see www.earthit.com).

Electronic signals from sensors are streamed to a logger, where the data are stored in digital format for subsequent downloading to a computer and processing in a GIS. A GIS is specialized computer software in which spatially referenced soil data are stored in layers, along with information on field boundaries and roads, and are linked to attribute data that describe each spatial feature in the GIS. Manipulation of the data allows a soil property, such as clay content or salinity, to be displayed as a map of high spatial resolution. Further details and examples of these techniques in action are given in chapter 2.

Techniques based on the electromagnetic (EM) spectrum can be applied remotely, as from aircraft, but the resolution is much lower than for proximal sensing and less relevant to soil survey for viticulture. For example, gamma-ray spectroscopy can be used proximally or remotely to measure gamma radiation emitted by naturally occurring radioactive elements in the top 30–45 cm of soil. Other methods make use of ground-penetrating radar (long-wavelength radiation) and laser-imaging radar in the ultraviolet, visible, and near-infrared regions of the EM spectrum.

In all cases, the sensor output must be converted to actual values of the soil property by calibration against laboratory-measured values, as in the case of clay content or salinity, or by using a model of soil chemical reactions, as in the case of estimating a soil's lime requirement (see chapter 3). After this is done, a large amount of spatially referenced data can be collected, stored, and used directly without the need for filtering through a soil classification. Data for the depth to a clay layer, soil texture and permeability, salinity, and so on are used to plan the layout of a new vineyard, including the best sites for individual varieties and irrigation design.

Even in established vineyards, precise knowledge of soil variation enables operations such as pruning, irrigation, fertilizing, and mulching to be carried out on a zonal basis, where a zone is an area in which the soil is less variable than in the vineyard as a whole. Note that the boundaries between zones in an established vineyard may or may not coincide with those between vineyard blocks. Because soil variation is often reflected in yield and fruit quality, it can be to the wine maker's advantage for these zones to be harvested separately (see chapter 6). Thus, the application of modern technology potentially closes the circle that began with the identification of particular *terroirs* in monastic vineyards on the slopes of the Côte d'Or centuries ago.

Regardless of whether grape growers and/or wine makers accept the concept of *terroir*, a knowledge of soil and its variability, and how it can be managed to achieve production objectives, is an essential component of vineyard management. Chapters 2 to 5 identify and describe relevant soil properties and associated management practices; chapter 6 concludes with a summary.

Summary

Soil is variable in the landscape because of the complex interaction of climate, parent material, organisms (plant and animal), topography, and time. As the parent material weathers in a variable environment, soil profiles develop with distinctive visual and behavioral properties that give rise to a range of soil types.

In Europe, most notably France, vignerons recognize a site's *terroir* as an expression of the varying environmental "endowments" of a site, in combination with human influences acting through the selection of grape varieties and viticultural practices. The distinctive character of a wine is said to reflect its *terroir* or "sense of place." However, modern methods for sensing soil properties remotely or close at hand at a high resolution (large scale), and storing this information in a GIS, provide a more precise description of the spatial structure of soil variability in a vineyard. Armed with this knowledge, grape growers-cum-wine makers can make better informed decisions regarding how to manage a vineyard to meet their objectives of fruit yield, wine style, and quality.

2

Site Selection and Soil Preparation

Terroir tests the limits of science, it is never precise, it is essentially an untellable story.

Edited extract, "The Age of Terroir," in *Daily Wine News* (December 2007), from Robert Geddes (2007), *A Good Nose & Great Legs: The Art of Wine from the Vine to the Table*

Soil: The Unseen Factor

Most people are unaware of the soil beneath their feet. Even in the home garden, attention is usually focused on the plants aboveground rather than the soil below. In the vineyard, growers ignore the soil at their peril because, left on its own, the grapevine can become a rampant weed. Indeed, in Roman times vines were allowed to climb over trees, which provided support, but made harvesting the fruit a hazardous task. Even today, native American vines such as *Vitis rupestris* and *V. riparia* can be seen festooning trees in their wild state.

In commercial viticulture, much emphasis is placed on canopy management to control the vigor of grapevines, especially when they are growing on deep, fertile soils with adequate rainfall (or supplementary irrigation). The book *Sunlight into Wine* by Smart and Robinson (1991) exemplifies this approach, and canopy management is regularly discussed in viticulture and wine magazines. Trellis systems have been devised to cope with vine vigor, and they provide optimum sun exposure and ripening conditions for the fruit. Bunch removal to control yield is common in French AOC vineyards (figure 2.1), and in the United States there are prototype machines designed to control vine growth and crop load through mechanical shoot thinning, leaf plucking, and bunch removal.

Less attention has been paid to the soil, even though vines depend on it to supply the two prerequisites for growth other than sunlight—water and nutrients. The grapevine is a hardy perennial that responds to the soil in various ways. For example, vine roots have been found at a depth of 5–6 m in the sandy, gravelly

Figure 2.1 Fruit dropped at veraison in a vineyard on the Terraces du Larzac, Coteaux du Languedoc, France.

soils of the Médoc region, France, and at similar depths in deep sandy soils in Sunraysia, Australia, and deep loess soils in Nebraska, USA. Even when water is limiting, a dryland (nonirrigated) vine can still develop an extensive root system because, like most plants, the vine adjusts its ratio of root to shoot to favor exploration of the soil for the limiting factor.

On the other hand, regularly irrigated vines such as those in many Australian vineyards on duplex soils (see chapter 1) have most of their roots in the top 40–60 cm of soil, which is the zone most affected by the irrigation water. Where there is a barrier to root penetration, such as a shallow layer of hard rock (e.g., calcrete in the Coonawarra region, South Australia), vine roots will extend laterally. These variations in root development influence the vine's vigor (expressed through its vegetative growth) and the yield and quality of its fruit. Thus, knowledge of the soil is a critical factor in the selection and preparation of new vineyard sites and the management of established vineyards.

Determining the Site

As outlined in chapter 1, "determining the site" in old established wine regions such as Burgundy, Tuscany, and the Rheingau has been achieved through centuries of practical experience. In the modern wine industry, determining the site starts by gathering as much data as possible on

- Climate
- Topography and geology
- Soil types and their distribution

- Availability of water for irrigation
- Potential for pest and disease attack

Climate

Climate ranks highly in site selection. Various climatic indices have been developed to identify the most favored regions for particular cultivars, ranging from the heat degree days (HDD) concept of California, to mean January temperature (MJT) for the southern hemisphere, to the global homoclime concept (box 2.1).

The probability of climate change should be considered with respect to site suitability for grapevines in general, and cultivars in particular. First, there is the overall warming trend, variously predicted from global circulation models to be in the range 0.4–2°C by 2030. Second, the models predict that some areas will warm faster than others, and rainfall patterns will change. In southern Australia, for

---

Box 2.1 Climate Indices for Site Selection

One of the earliest climate indices for wine grapes, developed in California by Amerine and Winkler (1944), is based on heat summation units called degree days. The heat degree days (or HDD) for a site for a 6- or 7-month growing season are calculated as follows:

HDD = Sum of [number of days $\times$ (mean daily temperature $-$ 10°C)]

The mean daily temperature is the average of the daily maximum and minimum temperature, which should be derived from records for at least 10 years. The 10°C (50°F) cutoff was chosen because few grape cultivars grow actively below this temperature, which is why the term growing degree days, synonymous with HDD, is used in some countries. Five macroclimatic regions are recognized, ranging from Very Cool (<1390 HDD, temperature in °C) to Very Warm/hot (>2220 HDD, temperature in °C).

Another system developed in Australia by Smart and Dry (1980) is based on the correlation between MJT and HDD for places of similar continentality. The continentality index of climate is measured by the difference between the mean January and mean July temperatures (for the southern hemisphere). More recently, Smart (2001) has promoted the homoclime concept, which is a way of comparing statistically the mean monthly temperatures and rainfall for a site with those for established wine regions around the world. For sites without a record of climate data, the climate can be estimated from national meteorological databases providing data on a 2 $\times$ 2-km grid. After the homoclime of a site has been established, the most suitable cultivars can be recommended, based on experience in other parts of the world. The risk of spring frosts and attack by fungal diseases such as downy and powdery mildew can also be assessed. Further details are given on the Web site www.smartvit.com.au.

example, it is predicted that winter rainfall will decline and summers will become hotter. Because the bulk of Australian wine is produced in hot inland regions, reliant on irrigation, there will be increasing constraints imposed by the availability of water. Also, cool-climate cultivars such as Pinot Noir and Riesling are likely to migrate southward. On the other hand, in Europe and North America, if sites that are currently risky because of spring frosts and winter freezes become warmer, the areas of wine grapes could move northward, or higher up hill and mountain slopes. These factors are important in any assessment of site suitability because investment in a vineyard has at least a 25-year time frame.

Site Indices

A more complete approach to site selection than that of climate analysis alone involves the development of a site index, combining knowledge of the requirements of particular cultivars with data for the factors listed in the previous section. Viticultural and soil experts have been asked to identify the key biophysical properties and weight them to form a composite index for sites in a region. For example, when applied to determining site suitability for wine grapes in West Gippsland, Victoria, soil, climate, and topography were weighted in the ratio 70%:20%:10% (reflecting the perceived order of their importance).

A similar approach was used in the hilly and mountainous terrain of central Virginia, USA—a region colder than West Gippsland. There, frost in spring and freezing in winter were serious constraints, so the most important factor in the index was altitude (affecting temperature). Altitude was weighted at 30%; followed by soil, 25%; existing land use, 20%; slope, 15%; and aspect, 10%. An example of a local site index (LSI) is that developed for Cabernet Sauvignon in Hawke's Bay, New Zealand—a region with a maritime climate of cool winters and warm to hot summers. In this index, soil properties such as maximum rooting depth, topsoil gravel content, and soil texture (affecting water-holding capacity and soil temperature) were combined with climatic variables such as growing season rainfall and mean October (spring) and January (summer) temperatures. The resultant index was better correlated with growth characteristics and fruit quality than several purely climatic indices. The LSI approach has been used to develop the concept of *digital terroirs*, which are described in chapter 6. A grower is more appreciative of such indices if the primary data are spatially defined using a GPS, entered into a GIS, and displayed on a colored map.

Although "determining the site" is only the first step in a chain of events in wine production, it is important because suitable land is becoming scarce in many of the more desirable wine regions. Also, the investment per hectare in establishing a new vineyard is large, so mistakes made in site selection can ultimately be costly. The remainder of this chapter focuses on the soil factor in site selection; it presents methods of gathering soil data and modifying the soil–vine environment to optimize vineyard performance.

Steps in Site Selection

Starting the Soil Survey

Chapter 1 introduces the role of soil survey in identifying soil variation. A soil survey allows soil limitations to vine growth to be identified and located, plans made to correct any problems, areas for particular varieties decided, and the irrigation layout to be designed, if required.

Commonly, a series of preliminary holes is "augered" to determine whether a full survey with sensing devices and profile pits is needed. If a full survey is to be done, the first step may involve remote sensing, such as with air photos or satellite images, or with EM techniques such as gamma radiation and laser-imaging radar. Stereoscopic analysis of air photos (giving a three-dimensional picture) has traditionally been used to map surface topography and vegetation, both of which can indicate soil changes.

An example of proximal sensing is an on-ground EM survey commonly performed with an EM38, as described in box 2.2. At its most basic, an EM-derived map shows the pattern of soil variation, which can be the result of a number of factors. If several soil properties are involved in determining the EM signal, the main value of the map is in indicating where to sample to cover the full range of soil variability, thus greatly improving the efficiency of sampling. Also, in contrast to a traditional soil map, the EM map shows the pattern of soil variation at a high resolution,[a] which in itself can be very helpful to vineyard managers.

In other cases, it is possible to correlate the EM data with the variation in a single property, such as soil water content or salinity or clay content, and produce a map of that property. Figure 2.2 is an example of a vineyard map in which the variation in the EM signal correlates well with the variation in soil clay content. In general, the accuracy of any EM-derived map of a particular soil property depends on how well the EM signal has been calibrated against actual measurements of that property in the field. The next section gives an example of calibration after an EM survey.

Calibration for Texture

Texture analysis is carried out for all soil surveys, whether they are based on a traditional 75 × 75-m grid, EM survey, or any other means of sensing. Texture analysis is especially important for obtaining estimates of readily available water (*RAW*), as discussed in chapter 4.

Calibration of an EM38 or similar instrument for texture is based on a particle size analysis of soil samples (the sample set) that represent the full range of texture variation in the vineyard. The samples are collected from specific points so

---

[a] High resolution means that the observation points are closely spaced.

Box 2.2 Electromagnetic Techniques for Soil Survey in Vineyards

The EM38 measures the bulk soil EC, which is affected by salinity, water content, clay type and content, bulk density, and soil temperature. However, if the in-field variation in one of these factors is large compared with the others, the EM38 can be calibrated to measure that dominant factor. The basic instrument is shown in figure B2.2.1A. It is normally placed in an insulated case to minimize temperature

(A)

(B)

Figure B2.2.1 (A) An EM38 in vertical orientation. (White, 2003) (B) An EM38 on a nonconducting sled being pulled through a vineyard. (Photo courtesy of Dr. Robert Bramley, CSIRO Sustainable Ecosystems, Adelaide, South Australia)

(continued)

Box 2.2 *(continued)*

variations and is mounted in front of a tractor or placed on a rubber or Perspex sled drawn by an all-terrain vehicle (figure B2.2.1B). A GPS mounted on the vehicle enables the position of the sensor to be tracked as it is moved over the land.

The EM38 works by EM induction, whereby a low-frequency, oscillating electric current in a transmitter generates small fluctuating currents in the soil, with their associated magnetic fields. The latter are detected by a receiver placed 1 m from the transmitter, as shown in figure B2.2.1A. The ratio of secondary to primary magnetic field strengths is directly proportional to the *EC* of a roughly cylindrical soil volume below. Placed upright, as shown in figure B2.2.1A, the effective measurement depth is about 1.5 m compared with 0.7 m when the EM38 is laid on its side (see www.geonics.com).

A related technique is that of measuring soil electrical resistivity (*ER*, the reciprocal of conductivity). A direct electric current is applied to electrodes built into the coulter wheels of a cultivator drawn behind an all-terrain vehicle. The potential difference between the electrodes is related to the soil's resistivity.

In Australia, the EM38 is used to measure vineyard soil variation, and the results have been calibrated against salinity, clay content (also reflecting water content), and soil depth (e.g., where a sharp change from soil to a rock layer occurs within the depth range of the instrument). When the EM or resistivity instruments are used in an existing vineyard, care should be taken that the signal is not distorted by steel posts and wires. This is not a problem for the EM38 when the row spacing is more than 2.5 m, provided the instrument follows the midline between rows, and it is less of a problem for *ER* measurements because the electrodes are closer together than the transmitter and receiver in an EM38.

that their EM reading and location are known. Ideally, soil salinity should be less than 0.6 dS/m because a higher salinity than this swamps the effect of any other soil variable on the EM signal. For particle size analysis, a known weight of dry soil is broken into its constituent mineral particles by mechanical and chemical treatment. Gravel and stones larger than 2 mm in diameter are excluded from the analysis, with the remaining material being called fine earth. Although the distribution of particle sizes in this fraction is continuous, for practical purposes the distribution is divided into classes. Not all countries use the same size/class divisions, but two of the most widely accepted classification schemes are as follows:

The International Scheme

Clay	<0.002 mm
Silt	0.002–0.02 mm
Fine sand	0.02–0.2 mm
Coarse sand	0.2–2 mm

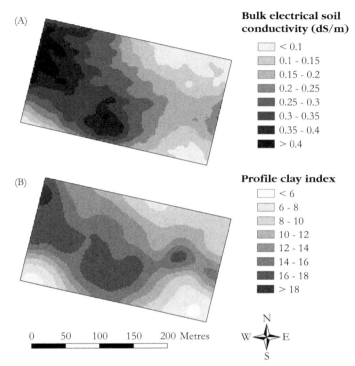

Figure 2.2 (A) EM38 map of a small vineyard in Sunraysia, Victoria. (B) A map of profile clay content derived from 130 sampling points in the same vineyard. The profile clay content is the mean of the clay content (measured as a percentage) at depths of 5–15 cm and 45–55 cm. (Courtesy of Dr. Robert Bramley, CSIRO Sustainable Ecosystems, Adelaide, South Australia)

The United States Department of Agriculture (USDA) Scheme

Clay	<0.002 mm
Silt	0.002–0.05 mm
Fine sand	0.05–0.1 mm
Medium sand	0.1–0.5 mm
Coarse sand	0.5–1 mm
Very coarse sand	1–2 mm

The amounts of clay, silt, fine sand, and so forth are calculated as a percentage of the fine earth, and their proportions determine the soil's texture. When these proportions are plotted on a triangular diagram, such as that shown in figure 2.3, we can pinpoint the textural class of the soil. For example, a soil with 35% clay,

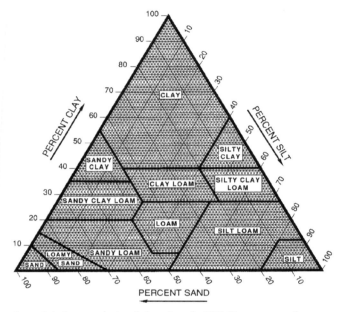

Figure 2.3 A textural triangle based on the U.S. Department of Agriculture particle size classification (similar to that used in Europe). (White, 2003; reprinted with permission of the United States Department of Agriculture)

Table 2.1 A Simplified Guide to Field Texturing

Sample preparation	Texture class	Distinguishing features
Take enough soil to fit into the palm of your hand, slowly moisten it with water and knead it between finger and thumb. Continue kneading and wetting until the soil "bolus" just sticks to your fingers and all aggregates are broken down (may take 1–2 minutes). Assess the texture class by feel, sound, and the length of the soil ribbon that can be squeezed between finger and thumb	Coarse sand	Particles are large enough to be seen and they grate together during kneading; bolus breaks apart[a]
	Fine sand	Particles may be difficult to see and feel, but can be heard grating when the bolus is kneaded close to the ear; bolus lacks cohesion
	Silt	Confers a smooth silkiness to the bolus, which is coherent but not very sticky (particles cannot be felt)
	Clay	The bolus is hard to knead initially but gradually becomes very sticky; forms an intact ribbon up to 50 mm long for a light clay and 75+ mm for a heavy clay[b]

[a] Humified organic matter makes a sandy texture feel smoother and more cohesive.

[b] The type of clay mineral affects the stickiness. When thoroughly wet, montmorillonite clay is sticky but kaolinite is not (see chapter 3).

30% silt, and 35% sand is classed as a clay loam. Experienced surveyors can estimate the approximate proportions of clay, silt, and sand by hand-texturing in the field, according to the guidelines given in table 2.1.

Based on the clay contents of the sample set, an EM calibration graph similar to that in figure 2.4 is drawn up. The line of best fit through the points is used to convert EM readings to clay content for the whole survey area and can also be applied to subsequent EM surveys done on this soil type.

Ground-truthing

Ground-truthing involves using soil analyses to check the accuracy of information derived from remote or proximal sensing. The samples for analysis are obtained from pits dug to more than 1.5 m in depth, or to parent material depth if that is less, or by augering or coring the soil. In the case of an intensive EM survey, such as that shown in figure 2.2, placement of the profile pits can be guided by the variation in EM signal to ensure the maximum range of soil variation is sampled. Depending on the number of pits dug in relation to the area surveyed, a more rigorous ground-truthing can then be achieved, which in turn improves the map's accuracy.

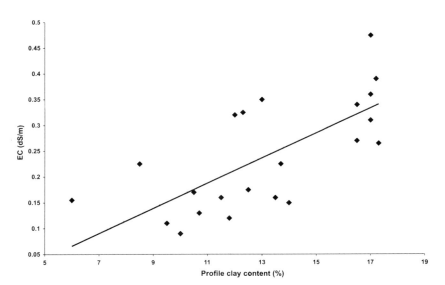

Figure 2.4 Calibration plot for electrical conductivity (*EC*) in terms of profile clay content. Twenty-five locations were selected at random from the EM38 map of figure 2.2. The line of best fit between *EC* (measured in deciSiemens per meter) and clay content (measured as a percentage) is shown. (Courtesy of Dr. Robert Bramley, CSIRO Sustainable Ecosystems, Adelaide, South Australia)

Table 2.2 **Precision and Accuracy in Soil Survey Measurements**

Precision	When a property such as soil depth is repeatedly measured in space or time and an average value (the mean) is calculated, precision refers to the uncertainty associated with that mean. The more often the depth is measured, the more certain the estimate of the mean (the value becomes more precise).
Accuracy	When an indirect method (e.g., an EM signal) is used to measure soil depth, the accuracy of the measurement depends on how well the surrogate variable (EM value) correlates with the actual depth measured by a standard method (with a meter rule).

The map's precision is checked by sampling with a soil auger to determine whether the location of changes in a soil property shown on the map agrees with that found on the ground. Indeed, soil augering is used to spot-check other soil properties, such as pH, color, stone content, and profile form, to confirm conclusions drawn from the relatively few profile pit observations. Table 2.2 summarizes the meanings of the terms "precision" and "accuracy," as applied in soil survey.

Describing a Soil Profile

When data from remote and/or proximal sensing are available, examination of soil profiles in backhoe pits is important not only for calibration purposes, but also for providing information on soil properties below the sensing depth. Horizons and layers in a soil profile were introduced in box 1.1, and they are distinguished by such obvious properties as color, texture, and structure. Descriptions of vineyard soils may refer to horizons or to depth intervals, such as 0–30 cm, 30–60 cm, 60–90 cm, and 90–120 cm. Whichever approach is used, the main focus is on identifying the limitations to successful vine growth, especially in respect to drainage and root extension, with the latter being more important in dryland vineyards. A survey report is enhanced by the inclusion of color diagrams of the main soil profiles annotated with a description of their important features, as listed in table 2.3. Table 2.3 draws on recommendations made by practicing soil surveyors such as Wetherby (2000), McKenzie (2000), and Maschmedt and coworkers (2002). Soil survey field handbooks, such as that of McDonald and colleagues (1990) and Soil Survey Division Staff (1993), give more details of methods.

To make measurements, a surveyor requires the following tool kit:

- A tape measure (preferably metric)
- A strong knife (and/or geological hammer)
- A pH test kit
- A Munsell color chart (or paint chips)
- Distilled water (or rainwater if not available) and plastic cup
- A hand lens (10× magnification)
- A small bottle of 5% hydrochloric acid (HCl) with dropper

Table 2.3 Soil Profile Properties Relevant to Site Selection for Vineyards

Key soil property	Rationale for selection	Method of measurement
Soil depth—whole profile and depth to a compacted (impeding) layer	Affects rooting depth and the possibility of waterlogging; may indicate a need for subsoiling and ripping	In centimeters; a compacted layer has a high resistance to penetration by a sharp, pointed object such as a penetrometer (see Consistence below)
Color	Orange to brick-red colors (resulting from Fe oxides) indicate good drainage; dark colors indicate organic matter; "mottling" indicates periodic waterlogging (see chapter 1)	By eye (specify whether soil is wet or dry); describe mottles and the background soil color separately
pH	Affects the availability of several nutrients; extremes of pH inhibit root growth	Universal indicator in the field or a pH meter in the laboratory
Texture, stones, and gravel	Influence a soil's water-holding capacity, drainage, structure, and ease of cultivation	Field texturing (table 2.1) or particle size analysis in the laboratory; proportions of gravel (2–200 mm) and stones (200–600 mm) estimated by eye
Structure, including sodicity and cracking	Influences soil stability when wet, hard-setting when dry; also aeration, drainage, and ease of root penetration; cracking clays swell when wet and shrink when dry	From a spade sample, assess degree of aggregation as well as aggregate sizes and shapes; note frequency and width of cracks; check stability of small aggregates placed in distilled water or rainwater (see chapter 4)
Consistence	Describes the strength and bulk coherence of a soil, which influence drainage, root penetration, and ease of cultivation	Test how easily dry 20–30-mm aggregates can be crushed between finger and thumb—"friable" if easy, "rigid" if very hard; measure bulk soil strength with a penetrometer
Presence of limestone or chalk ($CaCO_3$) and its hardness	Indicates neutral to alkaline pH; usually associated with good structure, but can restrict rooting if very hard	Drops of HCl cause fragments to fizz from CO_2 release; if rock, note whether fractured or not
Rippable rock or impeding subsoil layer	Influences rooting depth and soil drainability; if fractured, probably can be ripped	Assess hardness with a geological hammer; estimate the extent of fracturing

In addition to soil profile descriptions, growers are interested in the fertility status of the soils in their vineyards, which involves soil chemical testing (see chapter 3), and also whether there are potential biological problems (described later in this chapter).

Other Factors in Site Selection

Quantity of Water for Irrigation

A supply of good-quality water is essential for vineyards that need irrigation. The laws and regulations governing a grower's rights to underground and surface waters vary from country to country (e.g., how much can be used in any one season and the security of the water entitlement). In Australia, land and water resources are the responsibility of State governments, and the regulations vary among the States. Even within one State, the allocation of water can vary from region to region, depending on the amount of water available in each region in any one year and whether "carryover" of an unused allocation is permitted from one year to the next.

Wine grapes have a high value per megaliter (ML)[b] of water used, especially when a grower aims for a balance between fruit quality and yield through conservative irrigation practices. Thus, grape growers have the potential to be successful in buying water in a competitive market. To estimate the total water requirement for a vineyard a grower must consider three things:

1. The climate, especially the balance between rainfall and evaporation and the seasonal distribution of rainfall
2. The likely peak demand as determined by variety, planting density, yield objectives, and irrigation method
3. Additional factors, such as whether overhead sprinklers are to be used for frost control, or a winery is to operate onsite

The water may come from groundwater, surface water, reclaimed water, or a combination of these sources. Given that irrigation in most modern vineyards is applied through surface or subsurface drip lines, the likely peak demand will depend on the number of vines per hectare, and the number of drippers per vine and their delivery rate. Box 2.3 gives examples of how to calculate the peak delivery rate and total quantity of water required.

---

[b] One megaliter (ML) is equal to one million liters (10^6 L) and is approximately half the quantity of water in an Olympic-size swimming pool. It can be useful to equate irrigation to equivalent amounts of rainfall. In this case, 1 ML/ha is equivalent to 100 mm.

Box 2.3   Water Needs for an Irrigated Vineyard

Drippers for surface drip irrigation have delivery rates of 0.6–4 L/hour. Delivery rates of 0.8–1.2 L/hour are preferred for soils in which water entry is slow. In this case, two or three drippers might be placed in the drip line near each vine for the required amount of water to be applied during an irrigation period (figure B2.3.1).

Calculation of peak supply rate is illustrated by an example for midsummer (January) for the Riverland, South Australia. The vine density is 2020 vines per hectare, with two drippers per vine delivering 2 L/hour each. Thus, the pump and pipe system needs to deliver

$2 \times 2 \times 2020$ L/ha/hour = 8080 L/ha/hour

If the water required per vine, based on evaporative demand, is 16 L/day the irrigation system should run for 4 hr/day. Thus the daily water requirement is

$8080 \times 4 = 32{,}320$ L/ha $\cong 32{,}000$ L/ha

For one week's irrigation in midsummer, the water required per hectare would be

$32{,}000 \times 7 = 224{,}000$ L (0.224 ML)

The figures are comparable for the Griffith irrigation area in inland New South Wales for the same vine density.

Figure B2.3.1 Drippers in action in a vine row. Tensiometers are in place to measure soil water suction.

(continued)

Box 2.3 *(continued)*

A similar calculation can be done for the Lodi Appellation in the Central Valley in California (www.lodiwine.com). Peak demand occurs in July when a vineyard's average daily water need is 0.33 acre-inches. This amounts to 0.33 inches of water spread over 1 acre of land, and is equivalent to 8.4 L/m² (conversion factors from metric to U.S. units and vice versa are given in appendix 1). Thus, the total quantity of water required is

$8.4 \times 10,000$ L/ha/day $= 84,000$ L/ha/day

and one week's supply amounts to

$84,000 \times 7 = 588,000$ (0.588 ML)

Regulated deficit irrigation (see chapter 4) is now actively promoted in the Lodi Appellation with the aim of improving grape quality, so the peak water requirement, especially in the weeks up to veraison, would be 30%–65% of the previous amount, depending on the plant density and canopy management system. Thus, for low-density, vertically trained vineyards under regulated deficit irrigation at Lodi, the amount of water required per hectare per week could be as small as

$0.588 \times 0.3 = 0.176$ ML

or as large as 0.382 ML.

Note that if a grass cover crop is growing actively in the mid rows, the amount of water required could be twice as much as that required by the vines alone.

Water Quality

The most critical factor in water quality is the concentration of dissolved salts—mainly, sodium chloride (NaCl), calcium chloride ($CaCl_2$), and magnesium sulfate ($MgSO_4$). Over time, these salts accumulate in the soil because the water supplied by irrigation is used by plants and evaporated. The salt concentration is most readily measured by the *EC* of the water, usually expressed in units of deciSiemens (dS) per meter. From analyses of many water samples, the following approximate relationship has been developed:

1 (dS/m) = 640 mg/L total dissolved salts (*TDS*)

Table 2.4 explains the relationships between a number of the expressions used for the *EC* and salt concentration of water.

Because irrigation water is concentrated two- to threefold by evaporation from the soil, water of *EC* more than 0.8 dS/m should be avoided for vines, except where salt-tolerant rootstocks are used (see chapter 5). Less water is lost by

Table 2.4 **Common Expressions for the Electrical Conductivity or Salt Concentration of Water**

Expression	Units	Numerical equivalence
EC value	dS/m	—
EC value	mS/cm	Same as dS/m
EC unit	μS/cm	Multiply dS/m by 1000
Salt concentration	mg/L	—
Parts per million	μg/mL	Same as mg/L

evaporation with drip irrigation, so water of higher *EC* can be used than is acceptable for spray or flood/furrow irrigation. Subsurface drip irrigation is even more conservative of water than surface drippers.

Other useful analyses to be made on irrigation water are

- pH
- Na, Cl, and boron (B) concentrations, for potential toxicities and effects on wine quality
- Ca, Mg, and carbonate concentrations for assessing water "hardness" and for calculating the sodium adsorption ratio (see chapter 4)
- Iron (Fe) concentration, because if gelatinous Fe precipitates form, they can block drippers and even pipelines

Reclaimed Water

Growers may consider using reclaimed water when fresh water is in short supply. Reclaimed water is water from sewage works and processing industries that has been treated to a standard appropriate for its intended use. This type of water is being used successfully in wine regions that are reasonably close to a reliable source, such as in McLaren Vale, South Australia, and the Bellarine Peninsula, Victoria. Because reclaimed water is relatively high in nutrients (the water in McLaren Vale contains approximately 20, 9, and 24 kg/ha of nitrogen [N], phosphorus [P], and potassium [K], respectively, for every 100 mm applied), fertilizer inputs to the vines should be adjusted accordingly. However, the high *TDS* of reclaimed water (1000–1400 mg/L) can create problems after regular use unless the soil is leached by rain or a substantial application of low-salt irrigation water during winter, to remove the excess salts. One option is to blend or "shandy" reclaimed water with fresh water of low *TDS*. Large wineries also have the option of installing their own treatment plants to produce reclaimed water for the vineyard and other purposes (see chapter 5).

Soil Organisms

Some pest and disease organisms of grapevines live for part or all of their life cycle in the soil. The previous history of a site is especially relevant in indicating what tests should be carried out. For example, land previously under vines in an area that is not known to be phylloxera free should be tested for the root aphid phylloxera. Many vegetable crops such as tomatoes, potatoes, and carrots, as well as tree crops such as almonds, are hosts for nematodes that can attack and damage vine roots. Hence, the soil should be tested for the density and types of nematode present. Chapter 5 describes these organisms in more detail, as well as the soil-borne pathogenic fungi, bacteria, and viruses that attack grapevines.

Certain soil and vegetation factors predispose to biological problems. Phylloxera damage is generally more severe in clay and heavy clay soils, whereas nematode damage is more severe in sandy soils.

In testing for soil pests and disease, samples of moist soil should be collected by spade or auger from several locations, bulked, and sealed in polythene bags to be sent to a qualified laboratory for analysis. A bulk sample should represent no more than 4–5 ha. Testing for nematode types and numbers requires moist soil that is kept as cool as possible until analyzed. If unhealthy vines or other plants are seen onsite, separate samples should be taken where these occur, and in apparently healthy areas.

The presence or absence of phylloxera and/or nematodes will determine whether *V. vinifera* cultivars on "own roots" or on rootstocks should be planted. All *V. vinifera* cultivars are susceptible to phylloxera, as are any rootstocks that have *V. vinifera* stock in one parental line. Thus, although own-rooted cuttings are cheaper to buy, resistant rootstocks are the only sure way to avoid serious damage from phylloxera. Chapter 5 discusses the advantages and disadvantages of rootstocks. Regardless of whether rootstocks are chosen, the cuttings or "rootlings" (rooted cuttings) should be free from phylloxera, nematodes, and any potential fungal, bacterial, or viral pathogens (box 2.4).

Slope, Aspect, Nearness to Water, and Heat Input

Slope combined with aspect is important for maximizing the amount of solar radiation received by vineyards at higher latitudes (between 40 and 50°N or S). As shown in figure 4.15, chapter 4, part of the radiation absorbed at the land surface warms the soil. The sun's rays pass through a greater thickness of Earth's atmosphere at high latitudes than at the equator, which means the intensity of sunlight received by soil and vegetation is less at these high latitudes. The decrease in sunlight intensity is greater during autumn through winter into spring. However, because steep slopes that face toward the equator receive sunlight at a larger angle

---

Box 2.4 Preplanting Treatment of Rootlings and Cuttings

Hot-water treatment is essential to eliminate nematodes and phylloxera on the roots of dormant rootlings or to control infections of crown gall and phytoplasmas (which induce grapevine yellows) in dormant cuttings. The cuttings must be fully dormant before treatment. The treatment is as follows:

Rootstocks and *V. vinifera* cuttings   Hot water at 50°C for 30 minutes
Rootlings (bare roots)   Hot water at 50°C for 5 minutes

    Bundles of cuttings and rootlings should be hydrated and brought to ambient temperature before the hot-water treatment. Immediately after treatment, the bundles are plunged into clean water at ambient temperature for 20–30 minutes. Those that are to be kept in cold storage before being used may be dipped in a fungicide such as Chinosol (8-hydroxyquinoline sulfate) for control of pathogens.

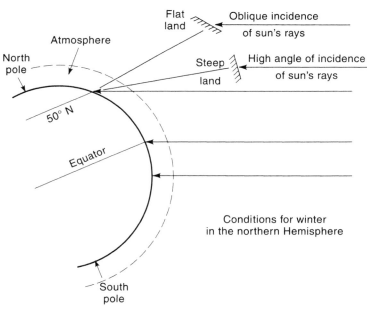

Figure 2.5 The influence of latitude and land slope on the amount of solar radiation received and absorbed at Earth's surface. (White, 2003)

of incidence than flat land, their potential for absorbing radiant energy and soil warming is greater. Figure 2.5 illustrates these effects.

    For these reasons, the most favored sites in cool-climate regions are predominantly south-facing slopes in the northern hemisphere and north-facing slopes in the southern hemisphere. However, this generalization needs to be modified for particular regions, such as in the Côte d'Or where, because of the local topography, many of the slopes are east facing. Unusually, in the cool Rheingau region of Germany, west-facing slopes are preferred because morning fogs are common during the critical autumn ripening period, and these usually clear by afternoon to the benefit of vines on west and southwestern slopes.

The duration and intensity of sunlight are not a problem in hotter climates, such as in La Mancha, Spain, the Central Valley of California, and the Murray Valley of Australia; however, excessive exposure of fruit to direct radiation can cause sunburn and a rapid rate of ripening. This problem can be managed through row orientation and row spacing, and by allowing more canopy shading. Row orientation and tree belts are important for protection from wind damage in some maritime areas (e.g., in many vineyard areas of New Zealand's North Island, and along the central and north coast of California).

Slopes also induce drainage of cold air, which helps to avoid damage to vines from low temperatures in winter and frost in spring. Most grape cultivars require at least 180 frost-free days, and frosts after bud burst can cause serious crop losses. In undulating topography, cold air drains from high points down slopes to valley bottoms, whereas warm air rises. In mountain regions such as in central Virginia, the mid slopes are the best for vines because cold air drainage makes the lower slopes and valley bottoms frost-susceptible in spring, whereas vines on the highest slopes are more likely to suffer winter freeze damage (at temperatures of $-2°C$ to $-15°C$), depending on acclimatization of the tissue.

The movement of air of different temperatures (and hence density) can make up a convection cell (figure 2.6). Such cells also develop near large bodies of water, such as lakes, large rivers, and estuaries. A large water body is a greater heat sink than nearby land, so warmer air over the water rises, to be replaced by cooler air draining from the land. The moderating effect of these convection cells means that

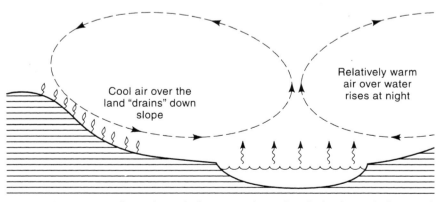

Figure 2.6 A convection cell on a clear night for a vineyard near a large body of water. (White, 2003)

vineyards near large water bodies do not suffer the extremes of temperature, either diurnal or seasonal, that occur in inland (continental) climates. Classic examples of this effect occur in the Médoc, which is bordered by the Gironde estuary to the east and the Bay of Biscay to the west, and the Margaret River region of Western Australia.

Modifying the Soil–Vine Environment

Site Potential

Each site has an inherent potential that reflects its suitability for particular varieties and the likely vigor (vegetative growth) of the vines to be grown. Given that site potential is determined by geology, soil, topography, climate, and human influence too, if the land form is reconstructed by earth moving, we can summarize the broad differences between low- and high-potential sites using a combination of the soil properties given in table 2.3 and additional geological factors (table 2.5).

Table 2.5 **Soil and Geological Features of Low- and High-Potential Sites**

Low-potential sites	High-potential sites
Often on conglomerates, sandstones, or shales and their metamorphic products, in which little "weatherable" mineral remains	Often on igneous rocks, especially basalt, diabase, and dolerite, relatively unweathered shales, their metamorphic products, and alluvium derived from these rocks
Underlying rock is hard and "unrippable"	Underlying rock is soft "saprolite" and rippable; no shallow barriers to root growth
Shallow soil (<0.5 m)	Deep soil (>1 m)
Sand to sandy loam, or a high proportion of gravel and stones; small water-holding capacity	Clay loam to light clay and <5% stones; large water-holding capacity
Low pH and exchangeable Al, or high pH and salinity[a]	pH 5.5–7.5 and no exchangeable Al or salinity
Little organic matter, usually a shallow topsoil and low mineral N supply	Plenty of organic matter, dark A1 horizon that can be >20 cm thick; potentially high mineralization of organic N[a]
Weak soil structure, especially in the subsoil; aggregates unstable in water[b]	Well-aggregated soil; firm aggregates stable in water
Poorly drained subsoil (mottled or blue-gray colors throughout)[c]	Well-drained soil profile (uniform orange to brick-red colors in subsoil and no mottling)

[a] See chapter 3.
[b] See chapter 4.
[c] See chapter 1.

In addition to this, we see topography influencing site potential through soil depth, drainage, temperature, and aspect. Soil on steep slopes is prone to erosion and therefore tends to be shallower than soil on gentle slopes and in valley bottoms. Soil at the top of slopes is usually better drained than soil at the bottom. Temperature and aspect effects are discussed earlier, but for a given suite of cultivars and scion–rootstock combinations, the interaction between rainfall and soil has a powerful influence on site potential, as expressed through vine vigor. Deep, well-drained soils that are naturally fertile and well watered will predispose to "excess vigor."

Excess vigor reflects a lack of balance between the vegetative and reproductive stages of a vine's development when too much vegetative growth occurs, especially between flowering and veraison, resulting in poor fruit set and shading that slows fruit ripening and lowers quality. Even in dry regions, irrigated vines on deep, fertile soils can suffer from excess vigor if too much water is supplied. Examples occur in the inland regions along the Murray and Murrumbidgee rivers in Australia and on alluvial soils of the Napa and Central Valleys of California (giving rise to the term "Californian sprawl"). Chapters 3 and 4 discuss further the problem of excess vigor induced by too much N and water, respectively.

Preparing the Site

Cultivation and Ripping

Soil treatment starts in summer. If the land has been cleared recently of trees or old vines, all residues must be removed or burnt and the soil cultivated or ripped to bring up old roots that can harbor pests and disease organisms. Deep cultivation to about 0.5 m in depth can be done with a moldboard plow, which turns a furrow slice. This is effective in bringing up old roots, but has the disadvantage of causing deep soil compaction, especially when the soil is wet, through pressure from the plowshare and tractor wheels in the furrow. In calcareous silty–sandy soils in Hérault, southern France, compaction of this kind has caused localized stunting and withering of vines, and yield decline (figure 2.7). Consequently, growers have turned to deep ripping (sometimes called subsoiling), which involves no soil inversion, and compaction under the tractor wheels is broken up by the ripper shafts that follow. However, old roots are not brought to the surface, so the soil may need to be fumigated to destroy nematodes (see "Preplanting" later in this chapter).

Deep ripping is often used to break up a compacted layer, either induced or naturally cemented, which prevents vine roots from penetrating deeply into the soil. For example, large rippers drawn by crawler tractors are used to rip soils in the Napa Valley in California; in the Western Cape wine region of South Africa a "finger mixed plow" is used to break up compacted B horizons in duplex soils without inverting the soil (figure 2.8). With this plow, as with the rippers, soil amendments

Figure 2.7 Stunted, unhealthy vines as a result of compaction in silty–sandy soils of the Hérault region of Languedoc–Rousillon, France.

Figure 2.8 A "finger mixed plow" used for soil preparation in vineyards of the Western Cape wine region, South Africa. The maximum depth of ripping is approximately 1.5 m. (Photo courtesy of Dr. Eduard Hoffman, University of Stellenbosch, South Africa)

Figure 2.9 A cultivator used for alleviating compaction in the mid rows of vineyards in the Western Cape wine region, South Africa. (Photo courtesy of Dr. Eduard Hoffman, University of Stellenbosch, South Africa)

such as lime can be incorporated at depth to raise the pH and improve soil structure, or gypsum alone may be added to improve structure.

In established vineyards, compaction from traffic in the mid rows can be serious, more so in sandy soils, but this, too, can be alleviated by ripping. South African scientists have developed a novel cultivator for deep disturbance of the midrow soil to alleviate compaction (figure 2.9). This implement also prunes the roots, which stimulates root growth and rejuvenates old vines of declining vigor. Many duplex soils in Australia and South Africa have compacted A2 horizons, which can be improved by mixing the A2 with the more humic A1 horizon.

Box 2.5 outlines the methods and soil conditions for successful ripping and deep cultivation.

Erosion Control

After deep cultivation or ripping, the soil surface is leveled to prepare for planting. Bare soil on slopes is vulnerable to erosion, especially when the vines are dormant in winter, so some form of erosion control should be used. On gentle to moderate slopes (0%–5%[c]), sow a cover crop that can be cultivated into the soil along the vine rows when the vines are to be planted. In this way, the soil between the rows is protected from erosion. Suitable cover crops are discussed in chapter 5. On steeper slopes (6%–15%), earthworks such as contour banks or terraces are required to break up long slopes and divert runoff water laterally, preferably to a grassed waterway at the edge of the block.

---

[c] A 1% slope has a vertical rise of 1 m over a distance of 100 m horizontally.

Box 2.5 Methods of Deep Ripping

To achieve shattering of a restricting layer or dense subsoil, deep ripping is best done in late summer–autumn. Ideally the soil water deficit (see chapter 4) should be about 50 mm in a sand or about 100 mm in a clay so that the soil does not smear, but tends to shatter, as the ripper blade passes through. In some cases it may be necessary to sow a winter cereal for harvest the following summer, to dry the soil profile sufficiently before deep ripping. Rip to a 1-m depth if possible. Achieving this depth requires a 1.5-m shank and a D8 or similar tractor. The subsoiler blade and its angle of strike are chosen according to the problem to be alleviated, as shown in table B2.5.1. If the soil is ripped at the correct water content (a friable consistence) at a 2-m spacing (with wings attached), the fissures created by the blade should intersect at the soil surface. This is important for good drainage.

Table B2.5.1 Recommendations for Ripping Subsoils

Soil problem	Blade type	Rake angle[a]
Hard rock	Do not rip	
Weathered rock	Rip with a wingless blade	
Cemented layer in profile	Rip with a wingless blade	
Clay subsoil	Rip with a winged blade	20 degrees from horizontal
Compacted sands	Rip with a 90-degrees point	

[a] The angle of the blade is critical for proper lift and shattering of the soil, with minimal lateral compaction

Adapted from Cass and coworkers (1998).

Old, established vineyards in Europe, such as the Hill of Hermitage in the Rhone Valley, France, have broad terraces on which vines are planted in rows up and down the slope (figure 2.10). In newer vineyards, the terraces are much narrower so that vine rows must follow the line of the terrace across the slope, as seen in the Napa Valley in California and the Friuli region of Italy (figure 2.11). Although the latter method provides better erosion control, mechanical operations in the vineyard are made more difficult. Generally, planting on slopes greater than 15% is not now recommended, although many of the old vineyards in the Mosel, Nahe, and Saar valleys in Germany have been planted on such steep slopes.

Preplanting

In land previously under vines, there are likely to be pests and disease that should be controlled before new vines are planted. For example, in parts of California, nematodes and oak root fungus are problems where old roots remain buried. Fumigation with methyl bromide (MeBr) has been used to correct these problems,

Figure 2.10 Old vines on broad terraces on the Hill of Hermitage, Rhone Valley, France.

Figure 2.11 Terraces in a newly established vineyard in the Friuli wine region, northeast Italy.

but because it is an ozone-depleting gas, MeBr use was phased out beginning 1 January 2005. Other fumigants such as Vapam (metam sodium) and 1,3 dichloropropene (Telone) have now been introduced. Fumigation is most effective when the soil is dry (providing air spaces for gas diffusion), the temperature is more than 12°C, and the surface is covered or packed down by rolling after the fumigant has been injected. However, because fumigants kill beneficial soil organisms as well as pests, the preferred control option is to use nematode-resistant rootstocks (see chapter 5).

Legume cover crops plowed in before planting stimulate beneficial soil organisms, helping young vines resist pest and disease attack until they are well established. Fertilizers can be applied when the cover crop is sown, but rock phosphate, lime, or gypsum should be incorporated along the rows when the soil is ripped because their effect is localized in the soil. If the topsoil is shallow, the soil may be mounded along the rows to increase topsoil depth and to improve drainage, taking care that when rootstocks are used the graft is not covered. Mounding around the vine trunks also provides protection against severe winter freezing.

Vineyard Layout

Irrigation Design

Irrigation is not used in most French AOC vineyards, and is used only sparingly or not at all by many producers of super- and ultrapremium wines in other countries. However, in Spain, Australia, and other countries where the summers are becoming hotter and drier, irrigation has become an essential part of vineyard management. Surface or subsurface drip irrigation is preferred because of its greater flexibility for varying the amount and timing of water application.

Vine water use can be made more efficient if water application takes account of such soil properties as depth, drainability, and water-holding capacity (see table 2.3). Although organic matter increases water-holding capacity, it is confined mainly to the topsoil, so its effect on the supply of available water in the whole profile is usually small. The influence of these factors on available water is discussed more fully in chapter 4.

Thus, if a soil survey reveals zones where one or more of these factors is substantially different, the irrigation design should be adjusted through the number of drippers per vine and their delivery rate (see box 2.3). Self-compensating drippers maintain pressure so that irrigation can be as uniform as possible between the start and end of a pipeline, and also on undulating ground. Keeping the pipe system as full as possible between irrigation sessions is important to reduce the filling time. Irrigation design is also influenced by vine spacing, both in row and between rows, as discussed later.

Drainage Design

One of the functions of deep ripping is to improve soil drainage. However, if poor drainage is likely as a result of relatively impermeable subsoil or a groundwater table that rises within 2 m of the soil surface, an artificial drainage system may need to be installed. Chapter 4 discusses the soil conditions predisposing to poor drainage and the means of alleviating this problem.

Vine Spacing

Vine spacing should be chosen according to site potential to satisfy the dual objective of highest quality fruit from each vine and optimum yield per hectare. Additionally, the vineyard should be easy to manage with the machinery available.

Although tradition plays an important part in determining vine spacing in different wine regions, some of the general principles distilled from research in several countries can be summarized as follows.

1. Low-potential sites need a closer vine spacing; high-potential sites
    need wider spacing. Vines are less vigorous in low-potential sites
    because of soil limitations, so close spacing can be used to increase
    production per hectare. The ultimate in close spacing—1 × 1 m,
    giving a plant density of 10,000 plants per hectare—is found in many
    of the top AOC vineyards on sandy–gravelly soils in the Médoc region
    (figure 2.12).

Figure 2.12 Vines at 1-m spacing on sandy–gravelly soil at Chateau Giscours, Margaux Appellation, Médoc, France.

2. In-row vine spacing has a more significant effect on yield and quality than between-row spacing because in-row vines should be far enough apart to provide cordon space for a balanced fruit load, without too much overlap of canes and shading. However, the vines should be close enough so that there is no unfilled cordon wire.

3. In addition to site potential, between-row spacing depends on the type of trellising used and the size of the machinery available for vineyard operations. Yield per hectare and fruit quality can be optimized with row spacings as wide as 3.5 m, and between-vine spacings up to 2 m, as is commonly found in high-production vineyards in Australia and the Central Valley, California. Books such as Smart and Robinson's (1991) *Sunlight into Wine* have detailed information on the types of trellis recommended for sites of different potential.

Summary

Site potential expresses the suitability of a site for particular grape varieties and the expectation of vine vigor displayed through vegetative growth. Soil is an important factor determining site potential, so detailed information on the soil's properties, spatially delineated through a soil survey, is an essential prerequisite for the establishment and subsequent management of a vineyard.

The basic aim of a soil survey is to identify and map the full range of soil variability at a site. The survey is greatly aided by remote and proximal sensing techniques, such as the use of an EM38 or similar instrument that responds to changes in the *EC* down to a depth of at least 50 cm. When the sensing positions are located with a GPS and the output is streamed to a GIS, a precise map showing soil variation can be produced. With such a map, points covering the full range of variation can be efficiently located on the ground and examined via soil pits. Samples from the pits can be used to calibrate maps for particular soil properties such as soil depth, clay content, available water capacity (*AWC*), and salinity.

A range of tests is recommended for soil chemical fertility (see chapter 3), physical condition (see chapter 4), and the presence of pest and disease organisms (see chapter 5). From these tests, a grower may decide on preplanting treatments such as deep ripping (to break up compacted layers), fumigation (for disease control), terracing (for erosion control), or sowing a cover crop. Soil amendments such as lime and gypsum, and rock phosphate fertilizer can be incorporated at this stage and a choice made about the use of rootstocks.

Although climate is important in site selection, it is considered here only through its interaction with topography, the moderating effect of nearness to water on temperature extremes, and soil temperature. If irrigation is to be used, the layout of the system should be based on the soil types present and their distribution. The quantity of water required at peak demand should be estimated and water quality assessed, with particular emphasis on *TDS* and elements such as Na, Cl, and B. The potential for salt leaching must be considered when reclaimed water of high *TDS* is used for irrigation.

3

The Nutrition of Grapevines

Technological wines that rely on dial-a-flavor yeasts
and additives to make up for dead soil don't taste
distinctive or real.

▨   From Robert Sinskey in *Daily Wine News* (March 2007)

The Essential Elements

Of the 115 known elements, there are 16 that grapevines must have to grow normally to flower and produce fruit. These essential elements, as shown in table 3.1,
are divided into

- Macronutrients, which are required in relatively large concentrations
- Micronutrients, which are required in smaller concentrations

Different ways of calculating nutrient concentrations in soil, plants, and liquid
are explained in box 3.1. Vines draw most of their nutrients from the soil, and so
table 3.1 also shows the common ionic form of each element in soil.

Ions are the charged forms of the elements that are released when compounds
of the elements are dissolved in water. For example, carbonic acid (H_2CO_3) is a
compound of carbon (C), hydrogen (H), and oxygen (O). In water, carbonic acid
dissociates into the ions H^+ and HCO_3^-. This is a chemical reaction that can be
written in shorthand form as

$$H_2CO_3 \leftrightarrow H^+ + HCO_3^-$$

The double arrow shows that the reaction can go either forward (to the right)
or backward (to the left), depending on the concentrations[a] of H^+ and HCO_3^-

---

[a] The concentration of H^+ ions in a solution at pH 5 is 10^{-5} M (molar), equivalent to one H^+ ion per
100,000 molecules of water, which is 10 times greater than the concentration of H^+ ions at pH 6.

Table 3.1 **Macro- and Micronutrients, Their Chemical Symbols, and Common Ionic Forms in Soil**

Macronutrient ($>$1000 mg/kg)[a]	Common ionic forms in soil	Micronutrient ($<$1000 mg/kg)[a]	Common ionic forms in soil
Carbon (C)	HCO_3^-, CO_3^{2-}	Iron (Fe)	Fe^{3+} (sometimes Fe^{2+})
Hydrogen (H)	H^+	Manganese (Mn)	Mn^{4+} (sometimes Mn^{2+})
Oxygen (O)	H_2O and many ions, (e.g., OH^-, NO_3^-, SO_4^{2-})	Zinc (Zn)	Zn^{2+}
Nitrogen (N)	NH_4^+, NO_3^-	Copper (Cu)	Cu^{2+}
Phosphorus (P)	$H_2PO_4^-$, HPO_4^{2-}	Boron (B)	H_3BO_3, $B(OH)_4^-$
Sulfur (S)	SO_4^{2-}	Molybdenum (Mo)	MoO_4^{2-}
Calcium (Ca)	Ca^{2+}		
Magnesium (Mg)	Mg^{2+}		
Potassium (K)	K^+		
Chlorine (Cl)	Cl^-		

[a] See box 3.1 for units of measurement.

Box 3.1 Units for Nutrient Concentrations and Amounts

Concentration (symbol C)[a] is the amount of a substance per unit volume or unit weight of soil, plant material, or liquid. For example, the concentration C of the element nitrogen (N) can be expressed as micrograms of N per gram of soil,[b] noting that

$$1 \ \mu g \ N/g = 1 \ mg \ N/kg = 1 \ part \ per \ million \ (ppm \ N) \qquad (B3.1.1)$$

An amount is the product of concentration and weight. For example, the total amount of N of concentration C (measured in micrograms per gram) in a soil sample of 100 g is

$$100C \ \mu g \ or \ 0.1C \ mg \qquad (B3.1.2)$$

Because all soil and plant materials contain some water, analyses are best expressed in terms of oven-dry (o.d.) weights. The o.d. weight is obtained by drying a soil sample to a constant weight at 105°C; for plant material the drying temperature is 70°C.

The amount of a nutrient is often expressed per hectare of vineyard. In this case, we need to calculate the weight of soil in 1 ha to a chosen depth, usually 0.15 m (15 cm). Unless it is measured directly, assume the bulk density for the top 0.15 m soil to be 1.33 Mg/m^3, so that the weight of dry soil per hectare is 1995 Mg or approximately 2000 Mg (2,000,000 kg). Thus, the conversion of a laboratory

(continued)

Box 3.1 *(continued)*

result of C_p (milligrams per kilogram) for "available" P in the top 0.15 m of soil to an amount per hectare of 0.15 m in depth becomes

$$C_p \text{ mg P/kg soil} \cong 2C_p \text{ kg P/ha.0.15 m} \tag{B3.1.3}$$

If the soil depth of interest is more than 0.15 m, as it would be for mineral N, the conversion factor is correspondingly greater.

Element concentration is sometimes expressed as a percentage of the dry weight. For example, a soil N concentration of C percent converts to an amount per hectare as

$$C\% = \frac{C\,\text{g}}{100\,\text{g soil}} = \frac{0.01\,C \text{ kg}}{\text{kg soil}} \tag{B3.1.4}$$

$$\text{Amount per hectare} = \frac{0.01\,C \text{ kg}}{\text{kg soil}} \times 2{,}000{,}000$$
$$= 20{,}000\,C \text{ kg N/ha.0.15m} \tag{B3.1.5}$$

Analyses of manures and biosolids are often quoted as a percentage of the element in the fresh or "wet" weight. To convert to a dry matter (DM) basis, divide the element concentration in the fresh weight (C_f percent) by the percentage of DM in the fresh weight as follows:

$$C\%(\text{DM}) = \frac{C_f}{100\,\text{g fresh weight}} \times \frac{100\,\text{g fresh weight}}{\text{g DM}} \times 100 \tag{B3.1.6}$$

[a] The concentration C of an element is an example of a variable. Variables are written in italics.

[b] Units of measurement and their abbreviations are given in appendix 1.

---

relative to H_2CO_3 and the strength of this acid. Carbonic acid is a weak acid because it does not completely dissociate in H^+ and HCO_3^- in solution.

Ions that carry one or more positive charges (+) are called cations and those carrying one or more negative charges (−) are called anions.

Sources of the Elements

In sunlight, vine leaves and green stems absorb C and O as the gas carbon dioxide (CO_2) from the atmosphere for the synthesis of sugars. Oxygen and H are also supplied in water (H_2O) derived from the atmosphere and the soil. Chlorine (Cl) is abundant as the Cl^- ion in the air and oceans. Sea spray containing Cl, sodium (Na), magnesium (Mg), calcium (Ca), and sulfur (S) forms an aerosol that is carried inland, where these elements are deposited in rain or as dust particles (dry deposition). Nitrogen (N) as dinitrogen (N_2) gas composes 78% of

Earth's atmosphere and can enter soil–plant systems through biological fixation, as described in chapter 5. Small amounts of mineral N are deposited in rain and as dry deposition. However, for the most part, the major source of the essential elements is a variety of minerals that decompose in the soil and parent rocks.

The macronutrients phosphorus (P), sulfur (S), Ca, Mg, potassium (K), and Cl are derived from rock weathering, as are the micronutrients iron (Fe), manganese (Mn), zinc (Zn), copper (Cu), boron (B), and molybdenum (Mo) (table 3.1). Additionally, rock weathering supplies other elements essential to animals, such as chromium (Cr), selenium (Se), iodine (I), and cobalt (Co); and yet others, such as arsenic (As), mercury (Hg), cadmium (Cd), lead (Pb), and nickel (Ni), which are toxic to humans and animals if they occur in too high a concentration.

Because the elements from Mn through to Ni normally occur at concentrations of less than 1000 mg/kg in soil, they are sometimes called trace elements (Fe is the one micronutrient that is not a trace element). However, all are potentially detrimental to vines if their concentrations exceed a threshold value, as determined by a competent authority (in Australia this authority is the Environment Protection and Heritage Council, www.environment.gov.au/about/councils/ephc/index.html). For example, the high Ni content of some soils formed on the rock serpentine at the northern end of Napa Valley, California, can cause problems for grapevines; there can also be too much Mg in serpentine, which is then out of balance with K.

Trace elements are deposited in dust from the atmosphere, especially after volcanic eruptions, and in sea spray. During the 20th century, however, additional releases from mining, industrial processing, motor vehicles, and agricultural chemicals have markedly increased trace element loadings in soil–plant systems via the atmosphere and through direct applications of solid or liquid waste, sometimes to unacceptably high levels.

Supplying Nutrient Ions to Vine Roots

The Vine Root System

As a perennial, the grapevine develops an extensive root system laterally and vertically when soil conditions are favorable. Observation of more than 200 root profiles in the United States showed that roots normally extend up to 1.5 m laterally from the trunk, and that about one fifth of the total root length is found below 1 m in depth. The thicker lateral roots in the top 30 cm of soil do not increase much in number after the third year from planting, but smaller "feeder" roots continue to grow horizontally and vertically from the main framework. Figure 3.1 shows healthy lateral vine roots in a well-drained soil.

Grapevines have evolved, in competition with trees, to explore a large soil volume at a low root density (expressed as the length of root per volume of soil).

Figure 3.1 Healthy lateral vine roots down to 50 cm in a well-drained red loam soil in the Barossa Valley, South Australia.

Little of the N, P, and S mineralized from organic matter is taken up from depths below 50 cm. However, some authors believe, without much evidence, that the distinctive character of wines produced from old vines is the result of an uptake of mineral nutrients such as Ca, Mg, Fe, and Mn from deep within the soil.

Although repeated cultivation of the mid rows inhibits root growth in the top 10–15 cm of soil, occasional cultivation will prune the roots and stimulate root growth. Where there is a subsoil obstruction, such as the dense B horizon of many duplex soils in southeastern Australia, less than 5% of vine roots may penetrate below 50 cm. Similarly, figure 3.2 shows that most of the roots in a soil formed on chalky limestone are confined to a shallow depth above the limestone. Deep rooting is also discouraged in irrigated vineyards if the irrigation strategy is to control water availability in the top 40–60 cm of soil only (see chapter 4).

The Absorbing Root

Vines absorb nutrients dissolved in the soil water in contact with the roots, which is called the soil solution. Thus, a nutrient such as N, which exists mainly in organic combination, must be released into the soil solution as an inorganic ion— ammonium (NH_4^+) or nitrate (NO_3^-)—to be absorbed by a root. The zone of active absorption is about 20 mm long, located a few millimeters back from the root tip,

Figure 3.2 Darkly stained fine roots confined to a shallow Terra Rossa over chalky limestone in the Barossa Valley, South Australia.

where there are root hairs (fine outgrowths from the epidermal cells), as shown in figure 3.3. Root hairs greatly increase the absorbing surface such that the uptake of a nutrient may exceed its rate of replenishment at the root surface, in which case the concentration around the root decreases and a depletion zone develops.

Depletion zones most commonly occur with the ionic forms of P, K, Fe, Zn, and Cu. These ions cross the depletion zone by diffusion in response to the concentration gradient between the root surface and soil farther away. Although this movement is slow, it is nonetheless important in maintaining the supply of these ions. On the other hand, the concentration of Ca, Mg, NO_3^-, Cl^-, and sulfate (SO_4^{2-}) in solution is normally large enough for root demand to be satisfied by the amounts swept along in water flowing to the root, and no depletion zone develops. Overall, ions such as NO_3^-, Cl^-, and SO_4^{2-} are considered to be "mobile" in soil whereas ions such as Ca^{2+}, Mg^{2+}, K^+, Fe^{3+}, Zn^{2+}, Cu^{2+}, and ortho-phosphate ($H_2PO_4^-$), which are mainly adsorbed on soil particles, are "immobile."

Mycorrhizas and Nutrient Uptake

Mycorrhizas are a symbiotic association between a fungus (the symbiont) and a plant root (the host). The fungus enhances the host plant's uptake of immobile nutrients such as P and, in return, obtains organic compounds for growth. The

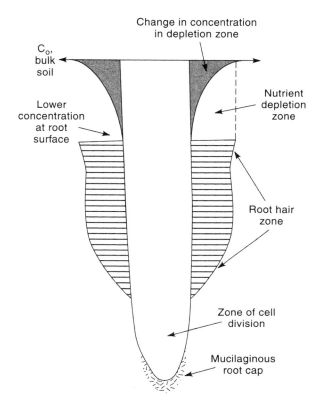

Figure 3.3 Diagram of a grapevine root showing root hairs and the nutrient depletion zone. (White, 2003)

most widespread type of mycorrhiza is an endomycorrhiza, so called because the fungus grows mainly inside the root where it develops branched structures called arbuscules that release nutrients inside the root cells (figure 3.4). This association is called an arbuscular mycorrhizal fungus (AMF).

Arbuscular mycorrhizal fungi are commonly found on *V. vinifera* roots and rootstocks. There are many strains of the fungus that survive as spores in the soil when no suitable host is present, and have no preference for a particular host. However, in a perennial plant such as *V. vinifera*, less effective strains may come to predominate over time, but midrow cover crops and weeds can counteract this trend by maintaining a more diverse AMF community in vineyard soils. Colonization by AMF is important for newly planted vines, especially in infertile soils.

The beneficial effect of AMF in enhancing P uptake is primarily through the fungal mycelium (see table 5.1, chapter 5), which provides a pathway for rapid P transport into the root, short-circuiting the depletion zone shown in figure 3.3. This could also be the mechanism for enhanced uptake of other immobile elements such as Cu, Zn, and K that is sometimes observed. An AMF may convert

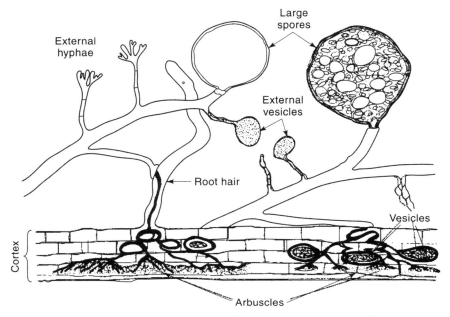

Figure 3.4 Diagram of an arbuscular mycorrhizal fungus infecting a vine root. (White, 2003)

organic P compounds to inorganic P and pass this P directly to the host. The mycorrhizal symbiosis is self-regulating in that, as soil P availability increases, the incidence of the infection declines. Mycorrhizal roots live longer than nonmycorrhizal ones and have an improved capacity to absorb water.

Nutrient Availability in the Soil

When considering availability to vines, it is helpful to divide the nutrients and other elements into groups as follows:

1. Macronutrient elements N, S, and P, which exist mainly in organic forms
2. Calcium, Mg, K, Na, and P, which bond to soil particles with varying strength (note that P is included in both groups 1 and 2)
3. Micronutrient elements Fe, Mn, Zn, Cu, and Mo, and trace elements Pb, Cd, Cr, and Ni, which form insoluble compounds and also bond strongly to mineral particles and organic matter
4. Chlorine and B (nutrients), and As (potentially toxic), which occur mainly in solution

Only a nutrient in the soil solution is immediately available to the vines. Assessing the availability of other nutrients that are distributed between solid and solution phases is difficult because both phases can contribute to the supply. Deficiency occurs when a nutrient's supply is insufficient to meet a vine's demand; conversely, when supply exceeds demand or a nonnutrient element accumulates in the root zone, toxicity may occur.

Availability of Nitrogen, Sulfur, and Phosphorus

About 99% of soil N is in an organic form not directly available to a vine. Thus, a soil with an organic C content of 1%, equivalent to 20,000 kg C/ha.15 cm,[b] will have between 1333 and 2000 kg N/ha.15 cm in organic form, depending on the C-to-N ratio of the organic matter (discussed later). The available forms (NH_4^+ and NO_3^- ions) are produced by mineralization—literally, the conversion of an organic form into an inorganic or "mineral" form. Mineralization also describes the conversion of organic S and P into inorganic S (SO_4^{2-}) and P ($H_2PO_4^-$ and HPO_4^{2-}). The reverse process of incorporation of C, N, P, and S into living microbial tissue is called immobilization. The role of soil microorganisms in the C cycle is discussed in chapter 5. Box 3.2 describes some consequences of N mineralization and nitrification.

The balance between mineralization and immobilization depends on whether there is enough of an element in an organic substrate to satisfy an organism's demand for growth. The demand for N, S, and P is directly linked to the demand for C, the main building material of microbial cells. For example, if the C-to-N ratio in the substrate is small, there is likely to be surplus N that is released as NH_4^+ and so net mineralization occurs. Conversely, if the C-to-N ratio is large, all the N is incorporated into microbial tissue and some mineral N may also be taken from the surrounding soil; net immobilization then occurs. Figure 3.5 shows that the critical C-to-N ratio for the tipping point between net mineralization and net immobilization is about 25.

As table 3.2 shows, materials such as cereal straw or vine prunings, which have high C-to-N ratios, are likely to induce net N immobilization in the soil. Low C-to-N ratio materials such as legume residues from a midrow cover crop should promote net N mineralization. Net mineralization also occurs when well-humified soil organic matter decomposes, but because its decomposition is slow, the release of mineral N is small compared with that released from fresh legume residues. However, some naturally fertile soils with more than 5% organic C can mineralize substantial amounts of N that may predispose to too much N uptake and excess vigor in grapevines.

---

[b] The expression "kg/ha.15 cm" means "kg per ha to 15 cm depth."

Box 3.2 Mineralization, Nitrification, and Their Effects on Soil pH

The first step in mineralization of organic nitrogen (N)—the production of NH_4^+ ions—is called ammonification. Because balancing OH^- ions are also produced, this step creates alkalinity.

Ammonium ions are taken up by vine roots, but much of the soil NH_4-N is consumed by specialist autotrophic bacteria that derive energy for growth by oxidizing NH_4^+ first to nitrite (NO_2^-) and then to nitrate (NO_3^-). This two-step process, carried out by *Nitrosomonas* and *Nitrobacter* species, respectively, is called nitrification. The net outcome of ammonification followed by nitrification is that 1 mole each of NO_3^- and H^+ are produced for every mole of organic N that is oxidized (see box 3.3).

Although nitrification is potentially acidifying, this potential is not realized if all the NO_3^- produced is taken up by plant roots, releasing HCO_3^- and OH^- ions in exchange, which cancel out the H^+ ions. However, NO_3^- is readily leached from the root zone by percolating water, in which case actual acidification takes place because the leached NO_3^- is accompanied by Ca^{2+} and other exchangeable cations that are displaced by the H^+ ions.

Soil acidification always occurs when an NH_4-N fertilizer such as ammonium sulfate is added to soil, because when the NH_4^+ ions are oxidized, two H^+ ions are released for every NO_3^- ion produced. The effect of this acidification is usually measured in terms of the "lime ($CaCO_3$) equivalent" required to neutralize the H^+ ions. Experience shows that the practical lime equivalent of $(NH_4)_2SO_4$ is about 5.4 kg pure $CaCO_3$/1 kg N in the fertilizer, and about 1.8:1 for NH_4NO_3 fertilizer.

The effect of urea fertilizer $((NH_2)_2CO)$ on soil pH is similar to that of soil organic-N compounds being oxidized, with one important difference. The soil pH around a granule of urea quickly increases to nine or more as ammonium hydroxide (NH_4OH) is formed from the urea. However, NH_4OH is inherently unstable and breaks down to release NH_3 gas and H^+ ions that neutralize the OH^-. The extent of any subsequent acidification then depends on how much of the remaining NH_4^+ is oxidized to NO_3^-, and how much of this NO_3^- is leached.

Similar principles apply to the mineralization of organic P and S compounds in soil, except that the critical C-to-P and C-to-S ratios are much higher than for N.

Availability of Calcium, Magnesium, Potassium, Sodium, and Phosphorus

The available forms of Ca, Mg, K, and Na are the cations Ca^{2+}, Mg^{2+}, K^+, and Na^+. These cations are attracted to negatively charged soil particles and can be exchanged by other cations in the soil solution, hence the name exchangeable cations. Although not a nutrient, Na^+ is included because its ions have an important effect on the behavior of soil colloidal particles in water. In addition to its organic

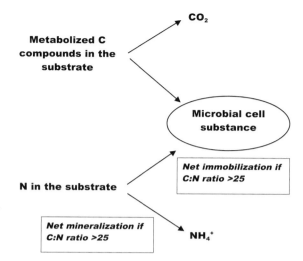

Figure 3.5 How the carbon-to-nitrogen (N) ratio affects the balance between net mineralization and immobilization of N in the soil. (Redrawn from White, 2003)

Table 3.2 Expected Net Nitrogen Mineralization from Soil Organic Matter or Returned Plant Residues

Organic material	C-to-N ratio	Net mineralization expected in 1 year
Well-humified soil organic matter	10–15	Yes, small
Legume residues (clover, medics, peas, beans)	15–25	Yes, significant
Grape leaves	30	A little
Cereal straw	40–120	No
Winter prunings	100	No

component, half or more of the P in soil can occur in an inorganic form that is strongly bonded to Fe and Al oxides.

Within a soil's clay fraction, identified in chapter 2, the negatively charged particles comprise crystalline clay minerals and organic matter. Iron and Al oxides are normally present as positively charged particles or as "coatings" on clay surfaces. All these components are important for nutrient retention in soil.

Retention of Nutrients by Clay Minerals and Oxides

Although clay minerals and oxides are of variable elemental composition, they exhibit several basic features:

1. Clay minerals have a platelike crystal structure with a large, flat surface area relative to the edge faces. Figure 3.6 shows the basic form of a clay mineral.

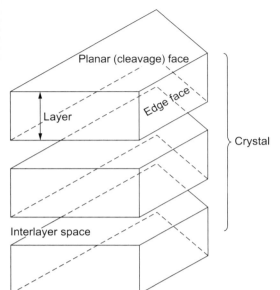

Figure 3.6 Diagram of the structure of a clay crystal. Such a crystal cannot be seen without a microscope.

2. There is some substitution of ions of similar size but different positive charge in the crystal structure, which results in a permanent negative charge in the particle. The most common substitutions are Al^{3+} for Si^{4+}, Mg^{2+} for Al^{3+}, and Fe^{2+} for Al^{3+}.

3. Cations in the soil solution are attracted mainly to the flat clay surfaces by the permanent negative charge. However, the much smaller edge faces also attract cations, provided the solution pH is above 6–7, depending on the type of clay mineral. At lower pH, the edge faces are positively charged and hence attract anions from the solution.

4. Like the clay mineral edges, Fe and Al oxides also show a variable charge that changes from positive to negative as the solution pH increases. However, up to about pH 8, the oxides are predominantly positively charged and thus are the main sites for the adsorption of anions such as $H_2PO_4^-$, Cl^-, NO_3^-, SO_4^{2-}, and some organic compounds. Of these, $H_2PO_4^-$ ions are the most strongly adsorbed because P has a strong affinity for Fe and Al, and bonds chemically to these atoms at the mineral surface. The term P fixation is used to describe this process.

Examples of the crystal structure of the common clay minerals are shown on the Web site www.virtual-museum.soils.wisc.edu/displays.html. Because they are very small (<2 μm), clay minerals have a large surface area per unit volume (called the specific surface area). The mineral montmorillonite, which belongs to

---

Box 3.3 Moles and Moles of Charge in Soil Chemistry

A mole of an element or compound is the molecular weight in grams, more commonly referred to as the molar mass. The standard against which all substances are measured is the stable isotope of carbon (^{12}C). On this scale, the molar mass of hydrogen (H), for example, is 1 g, calcium (Ca) is 40 g, and aluminum (Al) is 27 g.

The recommended unit of charged mass for cations, anions, and charged surfaces is the mole of charge, which is equal to the molar mass divided by the ionic charge. It follows that

- For H^+, because its ionic charge is $+1$, 1 mole of charge is $1/1 = 1$ g
- For Ca^{2+}, because its ionic charge is $+2$, 1 mole of charge is $40/2 = 20$ g
- For Al^{3+}, because its ionic charge is $+3$, 1 mole of charge is $27/3 = 9$ g

For clay minerals, oxides, and organic matter, the most appropriate unit is the centimole of charge $(+)$ or $(-)$ per kilogram (abbreviated cmol charge/kg). For example, the *CEC* of clay is expressed in cmol charge $(+)$ per kg because *CEC* is measured by the moles of cation charge adsorbed by the clay. The *CEC* expressed in this way is the same as that expressed in the obsolete units of milliequivalents (mEq) per 100 g.

---

the same group as the bentonite used to coagulate wine proteins, has the largest specific surface area. For example, if a heaped teaspoon (about 5 g) were spread out one layer thick, the clay particles would nearly cover a rugby football field.

The large specific surface area, coupled with the permanent negative charge, makes the clay minerals the site of many chemical and physical reactions in soil. The negative charge is measured by the number of moles of cation charge that can be held per kilogram of clay mineral. This is called the cation exchange capacity (*CEC*). Box 3.3 explains the concepts of moles and moles of charge, and table 3.3 summarizes some of the properties of the common clay minerals.

Retention of Nutrients by Soil Organic Matter

Soil organic matter has a heterogeneous composition not only because of the variety of organic residues deposited on the surface, but also because of the many biochemical processes, driven by microorganisms, which occur in the soil. The role of polysaccharides and polyuronide gums (sticky materials) in stabilizing soil structure is discussed in chapter 4. Here I describe the properties of charged compounds, such as polymerized phenols and organic acids, which form the structural "backbone" of humified organic matter.

As plant and animal residues decompose, a dark, humic material slowly forms, most easily seen beneath the litter from broadleaf trees or in well-made compost (figure 3.7). Humification leads to an increase in the number of organic

Table 3.3 **Some Features of the Common Clay Minerals**

Clay mineral type	*CEC* (cmol charge (+)/kg)	Specific surface area (m²/g)	Other salient features
Kaolinite	5–25	5–40	Strong bonding between crystal layers; interlayer spaces not accessed by cations or water; relatively large crystals and minimal swelling
Illite	20–40	100–200	K^+ ions fit snugly into holes between opposing crystal surfaces and hold the layers together; as K^+ is lost by weathering and replaced by partially hydrated Ca^{2+} and Mg^{2+}, the crystal expands slightly
Vermiculite	150–160	300–500	The end product of illite weathering in which K^+ ions have been replaced by Ca^{2+} and Mg^{2+}; limited swelling
Montmorillonite	100–120	750[a]	Layers are held in roughly parallel alignment when Ca^{2+} and Mg^{2+} are present; swelling increases as these ions are replaced by Na^+ and dispersion can occur (see chapter 4)

[a] For montmorillonite saturated with Na^+ ions.

Figure 3.7 Well-humified organic matter in compost. Note that plant parts are no longer recognizable.

chemical groups (designated as R) that can dissociate H^+ ions and become negatively charged, such as

$$R - COOH \leftrightarrow R - COO^- + H^+$$

and

$$\bigcirc - OH \leftrightarrow \bigcirc - O^- + H^+$$

These groups are weak acids of varying strength that gradually release H^+ as the pH changes from less than 3 to more than 8. Consequently, as the pH increases, the negative charge on the organic matter increases, and this translates into an increase in *CEC*. Compared on a unit weight basis, humified organic matter has 3–10 times the *CEC* of illite or montmorillonite clay at pH 7, so it makes a major contribution to the total *CEC* of any soil that is sandy and has more than 1%–2% organic C. Organic matter does not adsorb anions unless there are cationic impurities present such as Fe^{3+} and Al^{3+}.

Availability of the Micronutrients and Trace Elements

Iron, Mn, Cu, and Zn occur as metallic cations (see table 3.1), as do the trace metals Cd, Pb, and Ni (as Cd^{2+}, Pb^{2+}, and Ni^{2+}). The fully charged cationic forms exist only at low pH; as the pH increases, the metals hydrolyze and eventually precipitate as insoluble hydroxides. Box 3.4 describes this process. As a result of adsorption of the cations by clays and precipitation of hydroxides, the overall effect of an increase in pH is for the availability of these elements to decrease.

Molybdenum occurs as the anion molybdate (MoO_4^{2-}), which behaves like $H_2PO_4^-$. It is strongly adsorbed to Fe and Al oxides at low pH, and becomes more available as the pH increases. Boron occurs as neutral boric acid (H_3BO_3) up to pH 8, when it begins to dissociate and release the borate anion $B(OH)_4^-$, which is weakly adsorbed by oxides. Potentially toxic As and Se occur as the anion pairs arsenate–arsenite and selenate–selenite, respectively.

Metal cations also form complexes with organic compounds. Although these complexes are generally insoluble (when the organic matter is well humified), repeated applications of the metals can raise the concentration of the soluble fraction of the complexes to a point where they become toxic to beneficial soil organisms such as earthworms. This effect has been observed in old vineyards that have been repeatedly sprayed with Bordeaux mixture (a mixture of lime in copper sulfate ($CuSO_4$) solution) for the control of downy mildew. On the other hand, some organic compounds are used to form soluble complexes with Fe, Mn, Cu, and Zn to inhibit the formation of the insoluble hydroxides and make these nutrients more plant available (see box 3.4).

Box 3.4 Hydrolysis, Precipitation, and Organic Complex Formation

Hydrolysis is literally the splitting of a water molecule into its components: H^+ and OH^-. For example, the compound urea, found in the urine of grazing animals or applied as a fertilizer, hydrolyzes in moist soil. This reaction produces NH_4OH (a strong base) and H_2CO_3 (a weak acid). Because NH_4OH is more strongly dissociated than H_2CO_3, the result of hydrolysis is a greater concentration of OH^- than H^+ ions, and the pH increases, as discussed in box 3.2.

Another example is when micronutrient cations, such as Fe^{3+}, Cu^{2+}, Zn^{2+}, and Mn^{2+}, hydrolyze through the splitting of water molecules in their hydration shells. Iron exists as the hydrated cation $Fe(H_2O)_6^{3+}$ at pH less than 3 and hydrolyzes to the hydroxy form $Fe(OH)(H_2O)_5^{2+}$ as the pH increases. This hydrolysis continues with the stepwise release of H^+ ions until neutral $Fe(OH)_3$ is formed between pH 4 and 5. The other micronutrient cations and trace element metals (Pb, Ni, and Cd) behave similarly, with their hydroxides forming at higher pH. The hydroxides are all insoluble and precipitate from solution.

Specific organic compounds called chelating agents are used to modify the effect of pH on the solubility of micronutrient metals. Because of the high affinity of the organic compound for the metal, a chelate is formed that inhibits precipitation of the hydroxide. One example is the synthetic compound ethylenediamine tetraacetic acid (EDTA), which forms a complex $Fe\ EDTA^-$ that is more stable than $Fe(OH)_3$ up to pH 9. Similar chelation occurs between metal cations and natural organic compounds to form complexes that can be used as micronutrient fertilizers (see "Micronutrient Fertilizers" in this chapter).

Testing for Deficiencies and Toxicities in Grapevines

What You Can See

The first sign of a nutritional problem may be a "visual symptom"—an abnormality in growth, leaf color, flowering, or fruit set—that suggests one or more of the essential elements is in short supply, or perhaps there is a toxicity, as might occur with Na or Cl. Where the symptoms appear depends on an element's mobility. An element is mobile if it exists as an ion in the tissue (e.g., K^+, Na^+, and Cl^-) or because it is in high demand and is translocated from older, mature parts to young growing tissues (e.g., N and P). Some of the more obvious visual symptoms are

- Chlorosis, referring to leaf yellowing resulting from a lack of chlorophyll. Nitrogen deficiency typically causes overall chlorosis, whereas Fe, Mg, and Zn deficiencies cause yellowing between the leaf veins (interveinal chlorosis)
- Necrosis, referring to the death of leaf tissue. When deficiencies become severe, leaf and/or shoot tissue begins to die. Necrosis can also start as scorching of leaf margins, as in the case of Cl toxicity

- "Hen and chickens" appearance, which refers to bunches having a few normal berries and many small, seedless berries. This can occur with Zn and B deficiencies.

Table 3.4 summarizes nutrient deficiency symptoms and figure 3.8 shows examples of deficiencies. Goldspink and Howes (2001) give further details of deficiency and toxicity symptoms in grapevines.

Plant Analysis

Diagnosis of deficiencies and toxicities can be confusing when symptoms caused by different elements (e.g., Fe, Mg, and Mn) are similar. Also, visual symptoms appear only after the plant has suffered a check to growth as a result of a "hidden hunger" for the element, as illustrated by the relationship between fruit yield and a nutrient's concentration in vine leaves (figure 3.9). This relationship is the basis for diagnosing a vine's nutrient status by plant analysis or tissue testing. Figure 3.9 shows that as yield increases up to a maximum resulting from an increase in nutrient supply, the nutrient concentration in the tissue also increases. The degree of

Table 3.4 **Nutrient Mobility and Visual Symptoms of Deficiencies in Grapevines**

Nutrient element	Expression of symptoms
N	Mobile; overall reduction in growth and uniform leaf yellowing
P	Mobile; bronze coloration in older leaves early in the season
K	Mobile; starts as a yellowing of older leaf margins; as deficiency worsens, margins die and curl, and interveinal chlorosis develops
S	Mobile; deficiency rare in vineyards where S sprays are widely used to control powdery mildew; symptoms similar to N deficiency
Ca	Very immobile; shoot tips stunted and may die
Cl	Mobile; no deficiency seen because Cl is ubiquitous and readily absorbed
Fe	Immobile; young leaves show interveinal chlorosis; when severe, leaves may become very pale before developing necrotic spots
Mg, Mn	Mobile; interveinal chlorosis starts on older leaves and, when severe, necrosis extends inward from the leaf margins
Zn	Relatively immobile; stunted lateral shoots with small leaves showing "blotchy" interveinal chlorosis; fruit set may be affected; bunches show a "hen and chickens" appearance
Cu	Immobile; shortened internodes and death of shoot tips similar to Ca; flowering is also affected; rare in vineyards because of Cu sprays
B	Immobile; occurs in older leaves and may be confused with Zn deficiency because of "hen and chickens" appearance; the range between deficiency and toxicity is small
Mo	Immobile; necrosis of leaf margins; affects pollen tube growth and fertilization, resulting in poor fruit set

(A)

(B)

Figure 3.8 (A) Iron deficiency in a vineyard in the St. Emilion appellation, France. Note yellowing of the younger leaves and marked interveinal chlorosis. (B) Zinc deficiency in the Rueda region, Spain. Younger leaves are stunted and show interveinal chlorosis.

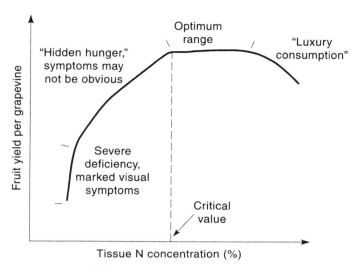

Figure 3.9 Stages from severe nutrient deficiency to luxury consumption according to tissue concentrations. (White, 2003)

deficiency can be assessed from where a plant's analysis lies on the approximately linear trend between severe deficiency and the optimum range.

Figure 3.9 also shows the critical value of nutrient concentration below which an increase in supply leads to increased yield. Note that in the "luxury consumption" range, yield may decline as a result of nutrient imbalances or outright toxicity. For example, too much N encourages vigorous growth that can cause excessive shading of basal buds, which leads to lower fruitfulness the following season. Excess N may also be a cause of early bunch stem necrosis. Boron toxicity may occur when vines are regularly irrigated with bore water.

Sampling Grapevines for Analysis

Which Part?

Because nutrients may be remobilized in response to a change in the site of greatest demand as a vine grows, the part to be sampled for analysis must be specified. It is generally recommended that leaf petioles be sampled because they are more responsive to a vine's nutrient status than leaf blades. Whether blades or petioles are sampled, washing and damp-drying are advisable to avoid the effect of dust on the analytical result.

The leaf sampled should be a basal leaf opposite a bunch—usually the fourth or fifth leaf from the base of a shoot, as shown in figure 3.10A. The petiole is separated from the blade and placed in a paper bag, to be sent to a laboratory for

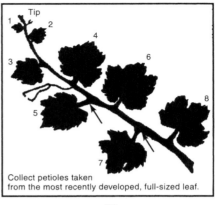

Select petiole opposite either
of the basal clusters during full bloom.

Collect petioles taken
from the most recently developed, full-sized leaf.

(A)   (B)

Figure 3.10 (A) Position of leaves to be sampled for tissue analysis (petioles) at flowering (bloom). (B) Sampling blades from mature leaves. (Christensen et al., 1978; reprinted with permission of the Division of Agriculture and Natural Resources, University of California, USA)

analysis as soon as possible. If toxicity is suspected, blades may also be sent for analysis because elements such as B accumulate more in the blade than the petiole. If blades are sampled, the most recently matured leaf on a shoot is recommended—usually the fifth to seventh leaf behind the tip (figure 3.10B).

Time of Sampling

The concentration of nutrients such as N, P, and K is greatest in young leaves and declines as the leaf ages. Traditionally, petioles are sampled at full flowering or bloom when the leaf is more sensitive to the external nutrient supply than at later stages of growth (box 3.5). The argument for even earlier sampling is that a grower has more time to respond with a remedial fertilizer application, but the disadvantage is that concentrations are changing more rapidly then. The exact time of sampling is not so critical if a regular sampling program (every year or second year) is followed (as described later).

For troubleshooting, or trying to confirm a deficiency or toxicity suggested by visual symptoms, leaves may be collected at any time. Leaves are best sampled early in the day when they are least likely to suffer water stress, and must not be sampled soon after any foliar spray has been applied.

Sampling Protocol

The traditional sampling method is to obtain a composite sample from a block of vines of the same variety on each soil type. Seventy-five to 100 petioles should be collected from randomly chosen vines in the block. Separate composite samples

---

Box 3.5 Stages in Grapevine Growth and Fruit Ripening

Bud burst   Bud dormancy is broken and young leaves begin to emerge and expand in spring (early October or April in cool-climate regions of the southern and northern hemispheres, respectively).

Flowering (or bloom in North America)   May extend over several weeks in cool climates, but the midpoint occurs when 50% of the fruit bunches (clusters) are in flower; cell division begins in the fertilized flower.

Fruit set   May extend over a few weeks, but the midpoint occurs when 50% of the bunches have berries 3–5 mm in diameter; all unfertilized berries should have been shed; from fruit set to veraison, berry growth is by cell division and expansion.

Veraison   Berries begin to soften and change color—a sign of ripening; mid-veraison occurs when 50% of the bunches are coloring (for red grapes); after veraison (postveraison), berry growth is mainly by cell expansion.

Maturity   Full ripeness attained; bunches are harvested.

Postharvest   Vines continue to grow, especially through root growth, provided temperature and soil water are adequate.

---

should be taken from areas that are performing badly or well. Label samples with the variety and vineyard block where the leaves were collected.

Even on one soil type, measurements at a very high density may show considerable variation in the nutrient concentrations over short distances. Figure 3.11 shows the variation in petiole P concentration in a Cabernet Sauvignon vineyard in the Coonawarra region, South Australia. Such variation probably reflects the influence of soil physical properties on growth, in addition to an effect of the soil's P supply. Thus, a strong case can be made for targeted sampling based on the known pattern of soil variability, which goes much further than the recommendation that separate samples be taken from "good" and "bad" areas. However, the cost of the extra analyses must be balanced against the potential value added by the more precise information, so methods of analyzing many samples rapidly and cheaply need to be deployed, as discussed in "Precision Viticulture" later.

Interpreting Plant Analyses

For a micronutrient such as Zn, results are reported in milligrams per kilogram dry matter, or as percent dry matter for a macronutrient (details are given in box 3.1). The result always has some uncertainty because of natural variability in the soil and vines, and because of analytical error. This uncertainty can be minimized with the best sampling and laboratory procedures, but too much emphasis should not be placed on a single set of results. Variations in weather from season to season will

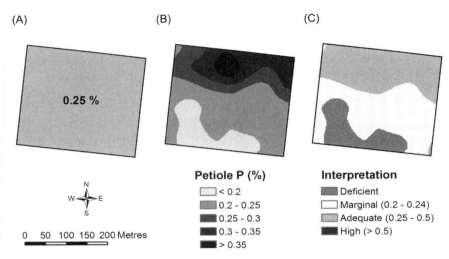

(A)   (B)   (C)

N
W—E
S

0   50   100 150 200 Metres

Petiole P (%)

☐ < 0.2
■ 0.2 - 0.25
■ 0.25 - 0.3
■ 0.3 - 0.35
■ > 0.35

Interpretation

■ Deficient
☐ Marginal (0.2 - 0.24)
▨ Adequate (0.25 - 0.5)
■ High (> 0.5)

Figure 3.11 (A–C) Maps of petiole phosphorus (P) concentration in a small Cabernet Sauvignon vineyard in the Coonawarra region, South Australia, assessed by the "industry standard" approach of analyzing a bulk sample (A) and intensive sampling of vines on a grid pattern (B). Although the same criteria are used to interpret maps A and B, the map (C) that is based on B, presents a different conclusion from A. (Courtesy of Dr. Robert Bramley, CSIRO Sustainable Ecosystems, Adelaide, South Australia)

affect leaf nutrient concentrations, so the best practice is to gather test results over several seasons to show trends in nutrient status. Any anomalous result will then become obvious, as shown in figure 3.12.

Record the condition of the vines and soil (especially soil moisture) at sampling time, as well as recent fertilizer or manure applications and cultural operations such as spraying with a fungicide. This information, together with the results of plant analyses and soil tests, provides a valuable record when compiled for individual blocks over several seasons and should be part of any integrated production system (IPS), as discussed later.

Using Critical Values

A critical value for each nutrient is ideally determined for an individual variety when growth is not limited by any other nutrient or the water supply (critical values for plants under water stress tend to be lower than for well-watered plants). However, varieties whether on own roots or rootstocks differ in their nutrient uptake and yield–nutrient relationships. This makes the task of calibrating plant analyses for commercial combinations of variety and rootstock so large that only a general set of values can be given, based on the much-studied Sultana variety (Thompson seedless). Tables 3.5 and 3.6 give the ranges of petiole concentrations for macronutrients and micronutrients, respectively.

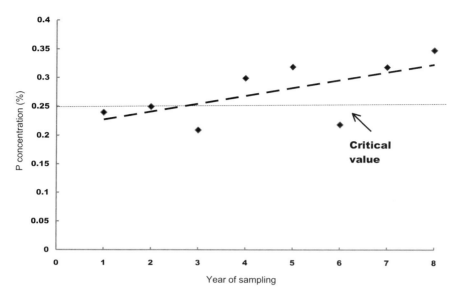

Figure 3.12 Example of a crop log for petiole phosphorus (P) analysis. Values in years 3 and 6 less than the critical value can be ignored because the general trend is upward.

Table 3.5 Macronutrient Concentrations in Petioles for Assessing the Nutrient Status of Grapevines

Macronutrient element	Very deficient (%)	Deficient (%)[a]	Low to marginal (%)[a]	Adequate (%)[a]	High to excessive (%)[a]
N (total)		<0.7	0.7–0.89	0.9–1.2	>1.2
N as NO_3^- (mg/kg)		<340	340–599	600–1500	>1500
P	<0.15	0.15–0.19	0.2–0.24	0.25–0.5	>0.5
P (Pinot Noir only)	<0.12	0.12–0.14	0.15–0.19	0.2–0.4	>0.4
K (with adequate N)		<1.0	1.0–1.7	1.8–3.0	>3.0
Ca		<1.2		1.2–2.5	
Mg		<0.3	0.3–0.4	>0.4	
Na					>0.5
Cl					>1.0

[a] Except where indicated otherwise.

Compiled from Robinson and coworkers (1997), Goldspink and Howes (2001), and Robinson (2005).

Assessment of a vine's N status is improved when NO_3-N concentration is used together with a total N analysis, but general leaf color and vine vigor should also be considered. There is an interaction between N and K in vines that results in petiole K concentrations being lower when vines are well supplied with N. For this reason, K analyses should always be interpreted in conjunction with N.

Table 3.6 **Micronutrient Concentrations in Petioles for Assessing the Nutrient Status of Grapevines**

Micronutrient element	Deficient (mg/kg)	Low to marginal (mg/kg)	Adequate (mg/kg)	High to excessive (mg/kg)
Fe		25–30	>30	
Cu	<3	3–5	6–11	>40[a]
Zn	<15	15–25	26–150	>450[a]
Mn	<20	20–29	30–60	>500[a]
B	<25	25–34	35–70	>100
Mo	<0.05	0.05–0.19	0.20–99	>100

[a] May indicate spray contamination.
Compiled from Robinson and coworkers (1997), Goldspink and Howes (2001), and Robinson (2005).

Soil Testing

Being a perennial, the grapevine is more suited to plant analysis than annual crops for which soil testing is more common. Soil testing involves analyzing the soil to assess its ability to supply a nutrient or to determine whether there is a potential toxicity. Although the analytical method is designed to extract the "available" part of the total element, this is an inexact concept because a vine's demand for a nutrient varies with its genetic potential, and many dynamic soil processes affect the supply of a nutrient to the root.

Apart from pH measurement, the most common soil tests are for available P and K. Soil testing for N is less common, mainly because the amount of available N (NH_4^+ and NO_3^-) is affected by several biological processes (see box 3.2). Many chemical extractants have been used, especially for P, so in interpreting a soil test the analytical method must be known. Whichever method is used, the range of test values corresponding to a very deficient, deficient, adequate, and excessive supply should be known from research trials conducted in vineyards. A critical value is then established, analogous to that shown in the graph of figure 3.9, except that "soil test value" is used on the horizontal axis. Calibration in this way is necessary to ensure soil test values are correctly interpreted, preferably by a specialist adviser. Also, for deep-rooted vines, the normal sampling depth of 0–10 cm or 0–15 cm may be inadequate. In Western Australia, for example, soil samples are normally collected in 10-cm intervals down to 50 cm.

Despite these caveats, soil testing is the only tool available at the time of vineyard establishment, when the whole soil profile should be sampled (see chapter 2). Soil tests are especially useful for pH and hence in assessing whether lime is needed, as described in box 3.6. Soil testing is also used to estimate "active" soil carbonate and hence Fe availability, and for indicating salinity and soil sodicity. In Western Australian vineyards, a soil test is recommended as a backup for

Box 3.6 Testing for pH and Lime Requirement

Soil pH indicates the likely effect of H^+ ions on plant growth and the potential for toxicities and deficiencies to occur. The pH stands for the negative logarithm of the H^+ ion concentration, measured in moles per liter of solution. The scale is from 0 to 14 pH units, with pH 7 being the neutral point (equal concentrations of H^+ and OH^- ions).

Soil pH is measured in the field with a universal indicator and color chart to an accuracy of ±0.5 pH units; alternatively, in a laboratory it is measured with a pH meter to an accuracy of ±0.05 pH units. Because soil is mixed with water (usually one part by weight of soil to five parts by volume of distilled water), the laboratory pH differs from the field pH. To minimize this difference, the soil can be shaken with a solution of 0.01 M $CaCl_2$ (1:5 ratio), which gives a pH value about 0.5–0.6 units lower than that in water and closer to the field pH. For this reason, it is important to know which method of measurement has been used when pH values are quoted.

The recommended pH ($CaCl_2$) for grapevines is between 5.5 and 7.5. Vines will grow outside this range, but are more likely to suffer a toxicity or deficiency. For example, the availability of Fe, Cu, Zn, and Mn decreases as the soil pH increases, for the reasons outlined in box 3.4. Conversely, at a pH ($CaCl_2$) less than 5.5, the hydrated cation $Al(H_2O)_6^{3+}$ begins to accumulate and can impair root growth and P uptake. In fact, clay in very acidic soils is dominated by these Al^{3+} ions that are released because H^+ ions attack the clay minerals and cause them to decompose slowly. Hydrolysis of the Al^{3+} ions produces a continuing

Table B3.6.1 Liming Materials for Vineyards

Material and its chemical composition	Neutralizing value of commercial grade (%)[a]	Comments
Burnt lime, CaO	>150	Reacts vigorously with water
Hydrated lime, Ca(OH)$_2$	120–135	Occurs as a very fine powder; difficult to handle
Dolomite, (Ca,Mg)CO$_3$	95–110	More soluble than CaCO$_3$; contains about 11% Mg
Limestone, CaCO$_3$ plus impurities	50–85	Neutralizing value depends on the concentration of impurities such as clay, silica, and Fe and Al oxides
Cement kiln dust, CaCO$_3$, CaO, K	90–110	Fine powder, by-product of cement manufacture; often pelleted; also supplies K

[a] Calculated relative to pure CaCO$_3$ as 100%.

Compiled from Goldspink and Howes (2001), and the Fertilizer Federation Industry of Australia (2006).

(continued)

Box 3.6 *(continued)*

supply of H^+ ions, which means that considerable amounts of lime are required to raise the pH to an acceptable value of 5.5 ($CaCl_2$) or more. A special laboratory soil test is used to measure a soil's "lime requirement," which ranges from 1 to 5 t $CaCO_3$/ha.

Laboratory analysis is also necessary to measure the neutralizing value (NV) of a liming material. Table B3.6.1 gives the NV of several such materials. Particle size is also important because the smaller the particles, the faster they dissolve in the soil. A high-quality liming material should have an NV more than 85% and greater than or equal to 60% of the particles less than 0.15 mm in diameter. Lime is most effective when cultivated into the soil.

---

plant analysis (e.g., extraction of soil in sodium bicarbonate is used together with petiole analysis to confirm the need for P).

Soil Testing for Salinity

Soil salinity is routinely measured using a conductivity meter that responds to the total concentration of dissolved salts in the soil (see table 2.4). A sieved sample of soil is made into a glistening paste with distilled water and the *EC* of the resultant solution—the saturation extract—is called the EC_e. In Australia, *EC* is commonly measured in a 1:5 soil-to-water extract so when soil *EC* values are reported, the method of measurement must be specified. Appendix 2 gives a table for converting from *EC*(1:5) to EC_e.

Chapter 2 discusses EM measurements for soil surveys, which can be calibrated to indicate soil salinity using a set of soil samples and a conductivity meter. Chapter 5, The Living Soil, discusses the tolerance of *V. vinifera* and rootstocks to salinity. The effect of Na^+ ions in creating soil sodicity is discussed in box 4.5, chapter 4.

Correcting Nutrient Deficiencies and Toxicities

There is a natural cycling of nutrients among vines, soil, and the atmosphere in which the grower can intervene at various points. This is best illustrated with reference to N, which is a key nutrient affecting canopy growth, fruit yield, must fermentation, and ultimately wine quality.

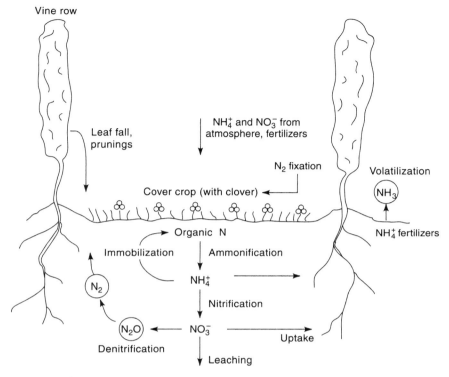

Figure 3.13 The nitrogen (N) cycle in a vineyard. (White, 2003)

Nitrogen Cycling

Figure 3.13 shows the N cycle in a vineyard. Atmospheric N_2 gas enters the cycle through legume N_2 fixation (see "Cover Crops and Mulches," chapter 5). There are also "free-living" bacteria in soil that can fix N_2, but their contribution in vineyards is small compared with that of legumes in cover crops.

The N-containing gases nitrous oxide (N_2O) and nitric oxide (NO) enter the atmosphere from natural or anthropogenic sources. Nitric oxide is oxidized in sunlight to nitrogen peroxide (NO_2), a major component of photochemical smog. Nitrogen peroxide dissolved in rain forms nitric acid, which contributes to "acid rain." Ammonia gas released to the air from NH_4-N fertilizers, animal dung, and urine (see box 3.2) can also be dissolved in rain or adsorbed as NH_4^+ ions on dust particles. Thus, NH_3 is eventually returned in rainfall or as dry deposition. These N inputs are most significant in industrialized areas where they can be as much as 60 kg N/ha/year, but inputs are as low as 5 kg N/ha/year in areas where emission controls are enforced, and in rural areas that do not have large animal feedlots or dairies.

Table 3.7 **Forms of Soluble Nitrogen Fertilizer Suitable for Vineyards**

Compound	Chemical formula	Nitrogen content (%)	Comments
Potassium nitrate	KNO_3	13	KNO_3 and $Ca(NO_3)_2$ are very
Calcium nitrate	$Ca(NO_3)_2$	15.5	soluble; suitable for fertiga- tion, but more expensive per kilogram N than other forms
Ammonium sulfate	$(NH_4)_2SO_4$	21	Supplies N and S in soluble forms; acidifying
Ammonium nitrate (Nitram)	NH_4NO_3	34	NO_3^- is immediately available; NH_4^+ is adsorbed in the soil and oxidized to NO_3^-
Monoammonium phosphate (MAP)	$NH_4H_2PO_4$	11	Water soluble; also supplies P; acidifying
Diammonium phosphate (DAP)	$(NH_4)_2HPO_4$	18	As for MAP
Urea	$(NH_2)_2CO$	46	Very soluble, prone to volatilize; acidifying; can be used as a foliar spray
Urea ammonium nitrate solution	$(NH_2)_2CO, NH_4NO_3$ in water	30–32	Has the advantages and disadvantages of urea and NH_4NO_3

Nitrogen Fertilizers

Table 3.7 gives the more important water-soluble N fertilizers used in viticulture. Fertilizers such as Nitram and urea supply only N whereas others such as KNO_3 and $(NH_4)_2SO_4$ supply more than one macronutrient and are called multinutrient fertilizers. Mixed fertilizers are made by mixing single or multinutrient fertilizers and are usually identified by their N-to-P-to-K ratio. For example, the mixed fertilizer Horticulture Special has a composition of 10–3–10, meaning 10% N, 3% P, and 10% K.[c]

Water-soluble N fertilizers are supplied as solids or liquids. Fertigation involves applying water-soluble N fertilizers in solution, usually by drip irrigation. There are also liquid organic products, such as seaweed and fish extracts, which have low N concentrations but are promoted on the benefits of their organic constituents. Applying such products as a foliar spray is an inefficient method of supplying N and other macronutrients because of losses through spray drift and incomplete

[c] In the trade, fertilizer analyses are sometimes given as the ratio of N to P_2O_5 to K_2O, but for simplicity all analyses are given as element percentages in this book.

absorption through the leaves. However, losses also occur when the more concentrated fertilizers are applied directly to the soil as solids or by fertigation.

Potential Losses from Nitrogen Fertilizers

The uptake of the fertilizer N applied in a single season may be less than 50%. Nitrate from $Ca(NO_3)_2$ and NH_4NO_3 is rapidly absorbed. Some of the NH_4-N from NH_4NO_3 and urea is immobilized by microorganisms, but most is oxidized to NO_3^-. Much of the NO_3-N that is not taken up by the vines or a midrow cover crop is lost by leaching or denitrification, the latter occurring when a soil becomes waterlogged (see "What Causes Poor Drainage?" chapter 4).

Not only is N lost as NO_3^- through leaching, but nitrification followed by leaching acidifies the soil (see box 3.2). Potentially the most acidifying N fertilizers are those in which all N is present as NH_4 N, such as in urea and $(NH_4)_2SO_4$. Less acidifying is NH_4NO_3, whereas $Ca(NO_3)_2$ and KNO_3 have no acidifying effect at all. This is one reason why $Ca(NO_3)_2$ is favored as an N fertilizer in vineyards, especially when applied by drip irrigation. Potassium nitrate is not recommended if the soil already supplies adequate K.

Urea applied to moist soil rapidly hydrolyzes to produce NH_4^+ ions and NH_3 gas that is lost by volatilization. High temperatures and wind accelerate NH_3 loss and, in extreme cases, up to 50% of the N in surface-applied urea can volatilize. Because the loss of NH_3 is most serious when there is insufficient rain or irrigation to wash urea granules into the soil, they are best placed a few centimeters in the soil or dissolved in irrigation water. However, if too much irrigation is applied, the urea may leach to greater depths before it hydrolyzes to NH_4^+ ions, in which case the acidity that develops through nitrification is difficult to correct because it is inaccessible to surface-applied lime.

Controlled-release fertilizer (CRF) or slow-release fertilizer (SRF) can be used to reduce N losses. Controlled-release fertilizers, such as sulfur-coated urea (SCU, 31%–38% N) and polymer-coated SCU (<30% N), are synthetic compounds for which nutrient release is slow and well controlled. The SRFs are synthetic or natural products for which nutrient release is slow, but not necessarily well controlled, because it depends on microbial action in the soil. Examples are blood and bone (5%–6% N) and hoof and horn meal (7%–16% N). Both CRFs and SRFs are sparingly soluble materials that do not produce the localized high concentrations of mineral N that predispose to N loss.

When to Apply Nitrogen Fertilizer

The timing and amounts of N influence vine vigor and canopy development, bud fruitfulness, juice N, and N storage in woody tissues. Figure 3.14 shows the periods of N uptake or remobilization within the vine in relation to its stage of growth. Root growth lags behind shoot growth from bud burst to the fourth or fifth leaf

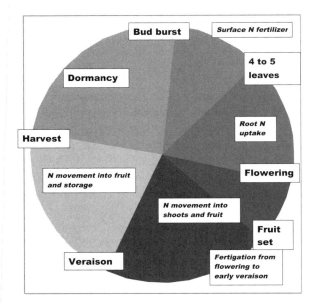

Figure 3.14 Grapevine nitrogen (N) uptake and fertilizer application in relation to physiological stages of growth and development. (Redrawn from Goldspink and Howes, 2001)

stage. During this period, N stored in the roots and trunk is mobilized and there is little uptake from the soil. After that stage, N uptake from the soil predominates as root growth accelerates to reach a peak midway between flowering and veraison. Thus, some N should be supplied to the vine at or soon after bud burst by surface application, because the soil is normally too moist for fertigation then. Depending on the vines' N status, more N fertilizer can be applied, preferably by fertigation, after flowering and up to veraison. An adequate N supply during this period has the most influence on juice N and, ultimately, wine quality.

Bunches are the main accumulators from veraison to harvest, with most of the N coming from remobilization and only a minor portion coming from the soil. Particularly in warmer regions, there is a postharvest period of root growth during which N is taken up from the soil. During this period, much N is also translocated from leaves and shoots to the roots and trunk, where it becomes a reserve for early shoot growth the following season, and has an important influence on subsequent vine fruitfulness. If the N status has been well maintained up to harvest, fertilizer is not needed postharvest when the risk of leaching loss by autumn and early winter rains is greater.

In summary, N applied at or soon after bud burst has a major effect on vine growth and crop load during the current season. Nitrogen supplied after fruit set has a major effect on berry N concentration, and on N storage in woody tissues for the following season. Normally, the fertilizer split between these times would be 1:1, but growers may need to adjust this according to experience. Table 3.8 summarizes recommendations to provide adequate N to vines with minimal N losses.

Table 3.8 **Recommendations to Minimize Nitrogen Losses in Vineyards**

Recommendation	Expected outcome
Do not apply soluble N fertilizers after harvest	Prevents N being leached by autumn and winter rains
Do not fertigate before flowering	Prevents leaching when soil is still wet from winter
Incorporate urea into the surface soil or wash in with irrigation water (but do not apply too much water before the urea has converted to NH_4^+ ions, which are not readily leached)	Minimizes NH_3 loss by volatilization, especially in warm to hot regions; avoids soil acidification at depth
Use a legume crop as green manure in the mid rows	Provides a slow release of mineral N to the vines as it decomposes; useful on sandy soil
Use a controlled-release fertilizer (CRF) or slow-release fertilizer (SRF) in row	Provides a slow release of mineral N; useful on sandy soil; some SRFs are acceptable for organic viticulture
Apply sufficient N to balance losses and removals in product (see "Balancing the Nitrogen Supply" in this chapter)	Provides adequate N for growth, but prevents excess mineral N building up in the soil

Balancing the Nitrogen Supply

The amount of N fertilizer required during one season depends on the net effect of N inputs (from soil mineralization, legume cover crops, and rain), N removed in harvested fruit and whether prunings are left in the vineyard, and any losses. Appendix 3 gives examples of how the N balance and N fertilizer required are calculated.

Adequate N is important not only for fruit yield, but also for maintaining an optimum N concentration in the berries for fermentation. Yeast strains vary in the efficiency with which they use combined and mineral N for growth. The critical measure of berry N for satisfactory fermentation and ultimately wine quality is the yeast assimilable nitrogen (YAN), comprising NH_4 N and amino acid N (excluding proline). Yeast assimilable nitrogen in the juice should be in the range of 200–480 mg N/L, with the optimum around 300 mg N/L, depending on the yeast strain. If YAN is too low, the fermentation rate is sluggish and may become "stuck." "Off" odors such as from hydrogen sulfide (H_2S) can be produced. Wine makers correct this problem by adding monoammonium phosphate or diammonium phosphate to the must during fermentation, but experiments in Western Australia suggest that N supplied in the vineyard produces wines of better flavor and aroma than those produced when N is added in the winery.

Conversely, if YAN is too high, fermentation is too rapid and poor-quality wine is produced, particularly in the case of red grapes. Residual protein in the must after fermentation causes haze in the wine. Especially when coupled with plentiful water, too high an N supply to the vine also leads to excess vigor, a topic to which we return in chapter 4.

Phosphorus Cycling and Phosphorus Fertilizers

As illustrated in figure 3.15, P cycling in a vineyard has similarities to N cycling with the following exceptions:

1. Phosphorus inputs from the atmosphere are negligible.
2. The chemistry of $H_2PO_4^-$ and HPO_4^{2-} ions is more complex than N because these ions are adsorbed onto clays and Fe and Al oxides, and P can also form insoluble precipitates. These processes are included in the term P fixation. One consequence of P fixation is that there is normally

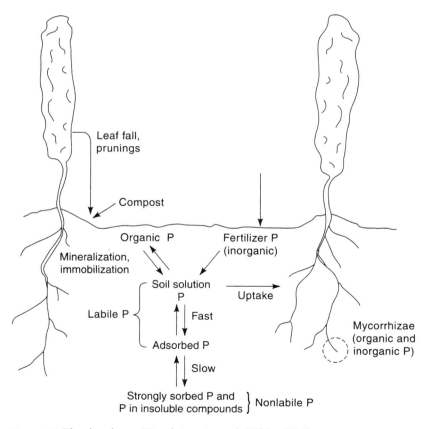

Figure 3.15 The phosphorus (P) cycle in a vineyard. (White, 2003)

little water-soluble P in the soil, except around dissolving fertilizer granules, and P does not readily leach from the soil in the way that NO_3^- does.
3. Because of P fixation and low water solubility, vines benefit from mycorrhizal symbioses, as discussed in "Mycorrhizas and Nutrient Uptake."

Soil testing for P aims to measure the amount of "labile" P, comprising soluble P and that fraction of solid-phase P that is readily released into solution. Water-soluble P fertilizers augment the labile P, but over time this P reverts to insoluble (nonlabile) forms. Thus, even though P is not lost by leaching or as a gas, fixation results in an ongoing need for P fertilizers or composted manures to maintain healthy vines, unless the soil is naturally rich in P minerals. Because of their large fixation capacity, soils such as red loams that contain much Fe and Al oxide require more P than other soils. Also, more P fertilizer is required at vineyard establishment than for the maintenance of mature vines. The maintenance application can be calculated from the fruit yield and the assumption that 0.6 kg P is removed per tonne of fruit, with an adjustment for P fixation that depends on soil type.

Table 3.9 shows the main forms of P fertilizer used in vineyards. Rock phosphates are of special interest because, being a natural product, they are acceptable in organic viticulture. A commercial rock phosphate should be "soft" rather than "hard," finely ground, and used only on soils of pH $(CaCl_2)$ of 5.5 or less to maximize its effectiveness. Because they are SRFs, rock phosphates are most effective

Table 3.9 Forms of Phosphorus Fertilizer Suitable for Vineyards

Compound	Chemical formula	Phosphorus content (%)	Comments
Single superphosphate (SSP)	$Ca(H_2PO_4)_2 \cdot H_2O$, $CaSO_4 \cdot 2H_2O$	9	80%–90% water soluble; remainder mostly soluble in neutral ammonium citrate; supplies P and S
Triple superphosphate (TSP)	$Ca(H_2PO_4)_2 \cdot H_2O$	19–21	80% water soluble; 15% citrate soluble; only traces of S
Monoammonium phosphate (MAP)	$NH_4H_2PO_4$	21–26	>90% water soluble, supplies N and P; suitable for fertigation
Diammonium phosphate (DAP)	$(NH_4)_2HPO_4$	20–23	80% water soluble and 20% citrate soluble, supplies N and P; suitable for fertigation
Rock phosphate	$Ca_{10}(PO_4)_6F_2$ with variable SiO_2, $CaCO_3$ and Fe, Al oxide impurities	6–18	Water insoluble and the citrate solubility depends on whether the rock is classed as "soft" or "hard"

if incorporated into the ripped soil at vineyard establishment. Phosphate ions are very immobile in all but sandy soils, so even water-soluble fertilizers such as superphosphate are more effective if mixed into the undervine soil by cultivation. Alternatively, soluble P can be supplied by fertigation.

Other Macronutrient Fertilizers for Vineyards

Calcium, Magnesium, and Potassium

Calcium is supplied in fertilizers such as $Ca(NO_3)_2$, single superphosphate, triple superphosphate, and the rock phosphates, as well as in various liming materials (see table B3.6.1). Gypsum also supplies Ca and, because it is more soluble than limestone, is useful for treating a subsoil acidity problem (caused by exchangeable Al^{3+}), because Ca^{2+} ions leached from the gypsum can displace the Al^{3+} ions, which then hydrolyze and precipitate as $Al(OH_3)$. Gypsum is also used to improve soil structure, especially in sodic soils, as discussed in chapter 4.

Dolomite is a limestone in which part of the Ca has been replaced by Mg. Magnesium can also be supplied as epsom salts ($MgSO_4$), which are soluble and can be applied as a foliar spray. Because many Australian soils have high exchangeable Mg^{2+} in the subsoil, an analysis of only the 0–10-cm depth may be misleading; petiole analysis is the best indicator of Mg deficiency.

Table 3.10 shows the common Mg and K fertilizers for vineyards. Potassium sulfate (K_2SO_4) is more expensive than potassium chloride (KCl), but is preferred in salt-affected soils in which Cl toxicity is a potential problem. Potassium sulfate, like KNO_3, has the advantage of supplying another major nutrient.

Table 3.10 **Common Magnesium and Potassium Fertilizers for Vineyards**

Compound	Formula	Element content (%)	Comments
Mg[a]			
Epsom salts	$MgSO_4 \cdot 7H_2O$	10	Water-soluble form of Mg fertilizer; used as a foliar spray at rates up to 10 g/L
K			
Sulfate of potash	K_2SO_4	41	Supplies K and S; preferred on saline soils
Potassium nitrate	KNO_3	38	Very soluble, supplies K and N; not to be used if vine N status is high; used as a foliar spray at rates up to 10 g/L
Muriate of potash	KCl	50	Very soluble in water; granular and easier to handle than K_2SO_4; not suitable as a foliar spray

[a] Magnesium is also supplied in some liming materials (see table B3.6.1).

During intermittent periods of low temperatures or under wet soil conditions before flowering, the basal leaves of vines may show symptoms of K deficiency. If petiole analysis shows the N-to-K ratio is too high and the leaves have a high NH_4 N concentration, the condition is called false K deficiency or, sometimes, spring fever. Although vines usually grow out of these symptoms as the season progresses, K fertilizer should be applied by fertigation if they persist.

Sulfur

Sulfur as elemental sulfur (S) is supplied in sprays for powdery mildew and as the SO_4^{2-} ion in fertilizers such as gypsum (14% S), single superphosphate (11% S), $(NH_4)_2SO_4$ (24% S), and K_2SO_4 (18% S). Additional sulfate S comes from the atmosphere in rain or as dry deposition, especially in industrialized regions. Spray S is supplied as a wettable powder with more than 80% of particle size less than 125 μm, and is easily mixed with water. Fine S particles are washed from leaves into the soil, but must be oxidized to SO_4^{2-} before they can be absorbed by vine roots. The oxidation of S is slow and produces some sulfuric acid (H_2SO_4) that contributes to soil acidification. With these S inputs, specific applications of S fertilizers are rarely required in vineyards.

Micronutrient Fertilizers

Micronutrients are commonly applied as foliar sprays, for the reason that elements such as Fe, Cu, Zn, and Mn are strongly bound by clays and organic matter and also form insoluble precipitates at neutral to alkaline pH (see box 3.4). For example, grapevines are susceptible to Fe deficiency in chalk and limestone soils, a condition called lime-induced chlorosis. Susceptibility to lime-induced chlorosis of vines on own roots or rootstocks is discussed in "Other Rootstock Attributes" in chapter 5.

Copper, Zn, and Mn are present in many fungicidal sprays, which, if used, remove the need for dedicated foliar applications. The effectiveness of a spray depends on how easily the nutrient penetrates a leaf and on its mobility within the tissue. A surfactant included in the spray improves leaf wetting and nutrient penetration.

Table 3.11 gives examples of inorganic salts or chelates that supply micronutrients. Natural chelates comprising Fe, Mn, or Zn complexed with a lignosulfonate are marketed on the basis of better absorption through leaves or from the soil than sulfates. Although this may be true for a soil application, it is not so for a foliar application because the larger chelated molecule is less easily absorbed through the leaf cuticle. Seaweed extracts containing micro- and macronutrients, vitamins, and amino acids are also marketed for use as sprays. They may provide

Table 3.11 **Examples of Micronutrient Fertilizers**

Fertilizer	Element content (% dry weight)	Comments
Borax	11% B	Soluble in hot water; apply in autumn to avoid phytotoxicity and to allow for translocation in the tissues
Copper sulfate	26% Cu	Soluble; also supplies some S; not required if Cu fungicidal sprays are used
Copper chelate	9%–13% Cu	Content varies depending on whether it is a natural or a synthetic chelate
Iron sulfate	20% Fe	Soluble, also supplies some S
Iron chelate	6%–10% Fe	Content varies depending on whether it is a natural or a synthetic chelate
Manganese sulfate	25% Mn	Soluble, also supplies some S
Zinc sulfate	22% Zn	Potentially phytotoxic if concentration is too high; "neutral Zn" ($ZnSO_4 \cdot Zn(OH)_2 \cdot CaSO_4$) is safer
Zinc chelate	14% Zn	May be a natural or synthetic chelate
Sodium molybdate	39% Mo	Soluble

Compiled from Goldspink and Howes (2001), and Christensen (2005).

benefits in addition to their nutrient content because of growth-promoting compounds, but this extra benefit is hard to quantify.

Because seaweed extracts and natural chelates cost more than the inorganic compounds, the latter are to be preferred provided that "spray-grade" forms are used (with negligible impurities that might cause leaf damage) and correct application procedures followed. Fifty percent of the nutrient may be lost through spray drift and 50% of the remainder may not be absorbed into the leaves. Although this is important in the short term, remember that residual nutrients on leaves will eventually reach the soil through rain wash and leaf fall, so in the longer term a spray-on micronutrient can build up in the soil.

Depending on vine density, recommended spray concentrations are in the range of 1–5 g/L in spray volumes of 200–1000 L/ha, giving application rates from 0.33 kg/ha for B to 1 kg/ha for Zn. Box 3.7 shows how to calculate the amount of either a macro- or micronutrient fertilizer to be applied per hectare or per vine row.

Organic Viticulture

What Does It Mean?

Organic viticulture aims to produce quality wine without the use of artificial fertilizers or synthetic chemicals. To the maximum extent possible, an organic system must operate as a closed system, with external inputs used only on an "as-needed"

Box 3.7 How to Calculate Fertilizer Application Rates

The amount of fertilizer required (the "rate") to supply a recommended nutrient is calculated as follows:

$$\text{Fertilizer rate (kg/ha)} = \frac{\text{kg nutrient per ha}}{\text{nutrient (\%) in the fertilizer}} \times 100 \qquad \text{(B3.7.1)}$$

So if the recommendation is for 50 kg N/ha and the fertilizer chosen is NH_4NO_3, the rate required is

$$\frac{50}{34} \times 100 = 147\,\text{kg/ha} \qquad \text{(B3.7.2)}$$

Rates for micronutrient fertilizers are calculated the same way, except that micronutrient rates are much lower.

Usually fertilizer is applied to the vine rows, in which case the amount required per row will depend on the row spacing. Suppose the recommended rate is y kg/ha and the row spacing is x m. The total length of rows per hectare will be $10,000/x$ and the amount of fertilizer required per meter of row is given by

$$\left(y \times \frac{x}{10000} \right) \text{kg} \qquad \text{(B3.7.3)}$$

The equation to convert from a fertilizer rate in kilograms per hectare to an amount per square meter is

$$\text{g fertilizer per m}^2 = \frac{\text{kg fertilizer per ha}}{10} \qquad \text{(B3.7.4)}$$

---

basis. The broad term "organic" includes biodynamic or biological systems, but biodynamic viticulture is recognized as having additional requirements to a simple organic system. These requirements include timing operations to coincide with phases of the moon and using special preparations (numbers 500–507) that are not fertilizers, but are intended to stimulate the soil's biological activity and root growth.

The organic viticulture movement, traditionally strong in France, is growing worldwide. A vineyard and its wine must be certified by an accrediting agency to be labeled organic. Internationally, certifying organizations include Ecocert International (www.ecocert.com), recognized in the European Union and by the National Organic Program of the USDA (www.ams.usda.gov), and the International Federation of Organic Agriculture Movements (www.ifoam.org).

In Australia, the peak body is the Organic Federation of Australia (www.ofa.org.au), which includes all sectors of the industry from growers to wholesalers, processors, certifiers, educators, and researchers. Members of this federation covering both organic and biodynamic products include the National Association for Sustainable Agriculture Australia (www.nasaa.com.au) and Australian Certified Organic (www.australianorganic.com.au), who apply the National Standard for Organic and Biodynamic Produce administered by the Australian Quarantine and Inspection Service (www.daff.gov.au). The Australian Biodynamic Agriculture Association (www.biodynamics.net.au) certifies only biodynamic products, which are then entitled to use the internationally recognized Demeter trademark (www.demeter.net). The first International Biodynamic Wine Forum was held in Beechworth, Australia, in 2004, and the movement is promoted by enthusiasts such as Max Allen, a well-known Australian wine writer (www.redwhiteandgreen.com.au).

For a vineyard to be certified as organic, it has to be established on organic principles from the start or it must undergo a 3-year conversion period, although after 2 years the wine can be labeled as "in conversion." Certified organic production means that grapes are grown without insecticides, herbicides, fungicides, and chemical fertilizers, other than those approved by the certifying agency, using cultural practices that minimize adverse effects on the soil and wider environment. Genetically modified plant material cannot be grown, nor can genetically modified yeast be used in fermentation. Chapter 6 includes a brief appraisal of the overall advantages and disadvantages of organic viticulture.

Complementary to the organic movement, growers in the major wine-producing countries are now encouraged to adopt an integrated production system (IPS), which is discussed in chapter 6. Briefly, IPSs involve minimizing pesticide and fungicide sprays through integrated pest management, which encourages a balance between natural predators and pests, and good canopy management to avoid disease problems. Fertilizer recommendations should be based on soil testing, and organic manures, compost, and mulch are used wherever possible. Midrow cover crops or permanent grass swards are advocated to improve soil structure and to provide a favorable habitat for beneficial organisms and the predators of pest insects.

Cultural Practices for Organic Viticulture

Weeds should be controlled by mechanical cultivation instead of herbicides. Mowings from a midrow cover crop can be thrown under the vines to serve as mulch. A cover crop containing legumes can be cultivated into the soil as a "green manure." The slow release of mineral N as the green manure decomposes minimizes the chance that any NO_3^- formed will be leached before the vine can absorb it.

Figure 3.16 An example of well-composted grape marc (pomace) in an organic vineyard.

Farm animal manures must be composted or be followed by at least two green manure crops after application. Biosolids derived from the composting of sewage sludge are generally prohibited unless specifically exempted by the certifying authority. The manure most commonly used in vineyards is poultry manure, followed by sheep manure. Composted manures should be pelleted to produce an odorless product in which N is at first stabilized by microbial immobilization, but the C-to-N ratio is reduced to less than 25 so that net N mineralization subsequently occurs in the soil (see figure 3.5). Composted manures and biosolids (if permitted) provide macronutrients such as N, P, K, and S; micronutrients; trace elements; and salts; so growers should take care that the amounts applied are not excessive for vines. This is particularly so for pig manure, which has relatively high concentrations of Cu, Zn, Mn, As, Cd, and Pb.

Composted green garden waste from cities is also used in vineyards, as is compost made from pomace (pressed skins and seeds from the winery, called marc in Australia; figure 3.16). Note that pomace may contain significant amounts of K that can be a problem for wine quality in some situations. For a general improvement in soil fertility and structure, composted manures and compost can be spread in the mid rows at rates up to 10 t/ha and worked into the soil in winter when the vines are dormant. For a more targeted application, compost is spread under the vines to a depth of at least 5 cm (a rate around 2 t/ha), serving as a mulch that suppresses weeds, reduces soil temperatures, and slows evaporation. Compost tea, an organic-rich liquid that drains from a compost heap, is sometimes applied as a foliar spray, but its benefits are not proved. Composting is discussed further in chapter 5.

Table 3.12 **Examples of Manures and Organic Fertilizers for Use in Organic Viticulture**

Material	N	P	K	Cu	Zn	Mn	Comments
Pelleted poultry manure[a]	3–4.6	1–2.8	1.2–1.7	30–170	220–300	300–580	Composted; variable depending on age and content of straw or sawdust
Blood and bone[a]	7–8	4–5					Slow release of N and P
Hoof and horn meal[a]	7–16						Slow release of N
Compost[a]	1.6–1.9	0.8–0.9	0.8–1.2	100–200	700–900	300–400	Variable depending on age and material used
Seaweed extract (Tasmanian Bull Kelp)[b]	0.75	0.15	1.2	6	30	62	Usually applied as a foliar spray; B is high at 70 mg/kg; contains plant growth hormones and amino acids

[a] Units for macronutrients (percent) and micronutrients (milligrams per kilogram) on a dry matter basis.
[b] Units for macronutrients (percent by weight); micronutrients (milligrams per kilogram); contains 15% water.
Compiled from Goldspink and Howes (2001), White (2003), and www.natrakelp.com.au

Table 3.12 lists the organic fertilizers acceptable in organic viticulture. Some are by-products from animal processing, such as hoof and horn meal, and others are plant-derived products such as seaweed extracts. As with biodynamic preparations, there may be nonspecific benefits from the stimulation of healthy biological activity in the soil. Table 3.13 lists the range of minerals from natural sources that are acceptable as fertilizers or soil amendments in organic viticulture.

Precision Viticulture

Precision viticulture embodies the concept of measuring soil variation with a dense spatial array of observations, and using this information to plan the layout, management, and harvesting of a vineyard most effectively. Figure 3.11 illustrated the point that, although the P status of a block may be deemed adequate based on one composite sample (figure 3.11A), the spatial pattern of variation shown in figure 3.11B reveals that more than half the vines are deficient or marginal in P. Correcting this deficiency would produce yield benefits, but the cost of sampling

Table 3.13 **Minerals from Natural Sources Acceptable as Fertilizers or Soil Amendments in Organic Viticulture**

Material	Function	Comments
Rock phosphate	Supplies P, Ca	Slow release; must be applied to an acidic soil to be effective; small liming effect
Elemental S	Supplies S	Wettable powder used for control of powdery mildew
Wood ash	Supplies K, Ca, and Mg	Timber must not have been chemically treated
Rock dust	Supplies Ca, Mg, some K, and micronutrients	Very slow release; must be finely ground
Rock potash (K_2SO_4)	Supplies K, S	Mined product
Epsom salts ($MgSO_4$)	Supplies Mg, S	Must not be chemically treated to improve solubility
Lime ($CaCO_3$)	Raises soil pH, supplies Ca	See box 3.6 for specifications
Gypsum ($CaSO_4 \cdot 2H_2O$)	Improves soil structure, supplies Ca	Must be from a natural source
Dolomite $(Ca,Mg)CO_3$	Raises soil pH, supplies Ca, Mg	See box 3.6 for specifications

Compiled from Goldspink and Howes (2001), and the National Standard for Organic and Biodynamic Produce (2005).

and analysis (190 samples in this case) by standard "wet" chemical methods may be prohibitive for a commercial vineyard.

The disadvantage of high cost can be overcome by the use of surrogate methods of analysis. The EM38 measurements described in chapter 2 are an example of a surrogate method for assessing spatial variation, which may be the result of variation in properties such as salinity, soil depth, or clay content. Midinfrared diffuse reflectance (MIR) spectroscopy, properly calibrated, is another example of a cheap and rapid surrogate method for measuring soil properties at a high spatial resolution. After calibration against a database of property values, measured by a standard method, MIR analysis can be applied to new sites. Figure 3.17 is an example of this approach. Alternatively, MIR can be calibrated using a "training" data set derived from a standard analysis of a limited number of samples, covering the full range of variability at a particular site. The surrogate method of analysis provides detailed spatial information on soil nutrient status, while substantially reducing the number of expensive analyses required (see also "Calibration for Texture," chapter 2).

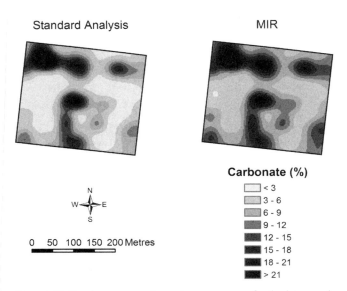

Standard Analysis   MIR

Carbonate (%)
    □ < 3
    ▨ 3 - 6
    ▨ 6 - 9
    ▨ 9 - 12
    ▨ 12 - 15
    ▨ 15 - 18
    ▨ 18 - 21
    ▨ > 21

0   50 100 150 200 Metres

Figure 3.17 Good agreement shown between maps of soil calcium carbonate (CaCO₃) content (5–15-cm depth) in a small Coonawarra vineyard, either measured by a standard method or predicted from midinfrared diffuse reflectance (MIR). The MIR was calibrated using a data set derived from the analysis of a large number of Australian soils. (Courtesy of Dr. Robert Bramley, CSIRO Sustainable Ecosystems, Adelaide, South Australia)

The yield response to soil variation can be mapped by acquiring data with a yield monitor attached to a mechanical harvester and importing it into a GIS. If the pattern of yield variation is consistent from year to year, blocks of vines can be identified for which soil management and fertilizer inputs are tailored to achieve desired outcomes, according to the style of wine produced and the price point. As more data are recorded from year to year under different weather conditions, the information for decision making can be revised so that uncertainty in the predicted outcomes is decreased. Precision nutrient management is potentially a big improvement on the traditional practice of taking bulk soil or plant samples for analysis, whereby variation within a block is "averaged." Specific nutrient management allows yield and grape quality to be optimized and potentially reduces the loss of nutrients such as NO₃⁻ through leaching and/or denitrification.

Tony Proffitt and co-authors (2006) give more details of the concept of precision viticulture and its applications in vineyards, a topic developed further here in chapter 6.

Summary

Grapevines rely on the soil to supply the following essential elements: N, P, S, Ca, Mg, K, Cl, Fe, Mn, Zn, Cu, B, and Mo. These elements are taken up as inorganic ions (nutrient ions) from the soil solution. Those from N through to Cl are called macronutrients because they are required in large concentrations compared with those from Fe to Mo, which are called micronutrients. Carbon and O are supplied as CO_2 through photosynthesis and H and O as water.

Most of the nutrient ions are retained in soil by adsorption onto charged clay, oxide, and organic matter surfaces, which affects their availability to vines. Elements such as P, Ca, Fe, Mn, Zn, and Cu can also form insoluble precipitates. Because of the strong adsorption of P by clays and oxides, and its tendency to form insoluble precipitates, "available" P in soil decreases markedly with time, a process called P fixation.

Although NH_4^+ and NO_3^- are the forms of N taken up by vines (the available N), 99% or more of soil N exists in an organic form. Similarly, a significant proportion of soil P and S is in organic forms, and the availability of N, P, and S is influenced by the balance between mineralization (conversion to a mineral form) and immobilization (incorporation into microorganisms and thence into humic compounds). Iron and the other micronutrients except B also form complexes of limited availability with organic matter.

Nutrient deficiencies and toxicities in vines can be assessed from visual symptoms, plant analysis, and soil testing. Protocols exist to determine which part of a vine to sample, when to sample it, how many samples to take, soil depth to sample, and so on. Regular plant analysis (every 1–2 years) is recommended to identify trends in nutrient status rather than relying on an analysis at one time. Spot samples can be taken at any time up to veraison if a problem suddenly arises.

Nutrient deficiencies are corrected through the use of fertilizers (as solids or dissolved in irrigation water), green manure (in the case of N), animal manures, and composts. Toxicity may really be a nutrient imbalance that can be corrected by adjustments to the fertilizer and/or manure regime, emphasizing the point that the nutrient content of added materials must be known. Another form of toxicity may arise when the soil pH is outside the optimum range of 5.5–7.5 (measured in $CaCl_2$ solution). Aluminum toxicity is a potential problem at low pH, whereas B and salt toxicity (resulting from too much NaCl) can occur at high pH. High soil pH, especially on limestone soils, can induce deficiencies of Fe, Mn, Zn, and Cu, but Mo becomes more available at high pH.

Precision viticulture provides the tools to determine, at a spatial resolution of 1–2 m, the distribution of any soil nutrient deficiencies and other limiting factors such as salinity. These spatial patterns can be matched against GIS-based maps of fruit yield or canopy size, acquired over several seasons, thereby allowing more effective application of fertilizers, irrigation water, and other forms of vineyard management.

Organic viticulture, including "biological" and "biodynamic" viticulture, aims to be a closed system with external inputs used only on an as-needed basis. Thus, synthetic herbicides, insecticide, fungicides, and chemical fertilizers are avoided, except where approved by a certifying body. Composted manures and vegetable matter are used when possible; obviously in most vineyards, the availability of such materials is limited.

4

Where the Vine Roots Live

I work the earth to put oxygen into it.

▓ Quote by Jean Michel Deiss, wine maker, in Andrew Jefford (2002),
The New France: A Complete Guide to Contemporary French Wine

Soil Structure

In chapter 3 we discussed how grapevines, being woody perennials, have the potential to develop extensive, deep root systems when soil conditions are favorable. One of the most important factors determining root growth is a soil's structure, the essential components of which are as follows:

- Spaces (collectively called porosity) through which roots grow, gases diffuse, and water flows readily
- Water storage and natural drainage following rain or irrigation
- Stable aggregation
- Strength that not only enables moist soil to bear the weight of machinery and resist compaction, but also influences the ease with which roots can push through the soil

The key concepts here are porosity, aeration and drainage, water storage, aggregation, and soil strength, each of which is discussed in turn.

Porosity

If it were not for the action of forces associated with the growth of plants, animals, and microorganisms, and physical forces associated with water and its movement, the elementary soil particles—clay, silt, and sand—would simply pack into an unconsolidated, disorganized heap. As a result of these forces, soil particles are organized into larger units called aggregates, between and within which there is a network of spaces called pores. Total soil porosity is defined by the ratio

$$\text{Porosity} = \frac{\text{Volume of pores}}{\text{Volume of soil}}$$

A soil's A horizon, containing organic matter, typically has a porosity between 0.5 and 0.6 m^3/m^3 (also expressed as 50%–60%). In subsoils, where there is little organic matter and usually more clay, the porosity is typically 40%–50%. Box 4.1 describes a simple way of estimating soil porosity.

Box 4.1 Estimating Soil Porosity

A simple equation for porosity, expressed as ϕ (phi), is

$$\phi = 1 - \frac{\rho_b}{\rho_p} \tag{B4.1.1}$$

In this equation, ρ_p (rho p) is the average density of the soil particles, assumed to be 2.65 Mg/m^3. The term ρ_b (rho b) is the soil's bulk density, which ranges from less than 1 Mg/m^3 for soils rich in organic matter to 1.0–1.4 Mg/m^3 for well-aggregated loamy soils, and to 1.2–1.8 Mg/m^3 for sands and compacted subsoils. Thus, for a loamy soil with a ρ_b value of 1.33 Mg/m^3, we have

$$\phi = 1 - \frac{1.33}{2.65} = 0.5 \tag{B4.1.2}$$

giving a porosity of 0.5 m^3/m^3 or 50%.

To measure a soil's bulk density, take five to six intact cores with steel cylinders, preferably at least 6 cm in diameter and 6–10 cm deep. Trim the soil so that the dimensions of the soil core are the same as the cylinder and the soil volume is easily calculated. Dry the cores in an oven at 105°C and weigh each core to obtain the weight of oven-dry (o.d.) soil. The bulk density ρ_b of each core is calculated from the equation

$$\rho_b = \frac{\text{Weight of o.d. soil in the core}}{\text{Volume of soil core}} \tag{B4.1.3}$$

Calculate the average value of ρ_b for the several samples taken. Note that soil bulk density in the rows and mid rows is likely to be different because of compaction by wheeled traffic in the latter.

Total porosity is important because it determines how much of the soil volume roots and water can occupy. Equally important are the shape and size of the pores. The pores created by burrowing earthworms, plant roots, and fungal hyphae are roughly cylindrical, whereas those created by alternate wetting and drying appear as cracks (figure 4.1). Overall, however, we express pore size in terms of diameter (equivalent to a width for cracks). Table 4.1 gives a classification of pore size based on function.

Figure 4.1 A vineyard soil with swelling clays showing cracks on drying.

Table 4.1 Relationship among Pore Size, Formative Forces, and Function

Pore diameter (μm)	Biotic or physical agent	Pore function
5000–500	Cracks resulting from drying, earthworms, primary plant roots	Aeration and rapid drainage
500–30	Grass roots, small insects and animals	Normal drainage and aeration
30–0.2	Fine lateral roots, fungal hyphae, and root hairs	Storage of "available water" (see "Aeration and Drainage" in this chapter)
<0.2	Swell–shrink water associated with clay minerals	Retention of residual or "nonavailable" water

Compiled from Cass et al. (1993) and White (2006).

Aeration and Drainage

Pore space is normally occupied by water (the soil solution) and air. When all the pores are filled with water, the soil is saturated. As water drains out or evaporates from the surface, air enters the pore space and the soil becomes unsaturated. Air enters the largest pores first, primarily by mass flow, followed by diffusion into the smaller pores. Normal respiration of soil organisms and plant roots consumes oxygen (O_2) and produces carbon dioxide (CO_2), and the exchange of these gases between the atmosphere and soil air is called aeration.

Aeration and drainage depend on at least some of the pores being continuous. As a result of gas exchange through these pores, the concentration of dinitrogen (N_2) and O_2 in the soil air is similar to that of the atmosphere at 78% and 20% (by volume), respectively, whereas the CO_2 concentration ranges from 0.1% to 1%. However, in most soils there are also "dead-end" pores (up to 5% of the total) that remain gas filled even when the soil appears to be saturated. The O_2 in these pores is soon depleted, with the result that respiration switches from aerobic (in the presence of O_2) to anaerobic (in the absence of O_2), and gases such as methane (CH_4) and nitrous oxide (N_2O) are produced.

Consistent with total porosity, the air-filled porosity is expressed as a fraction of the soil volume (units of cubic meter per cubic meter), according to the ratio

$$\text{Air-filled porosity} = \frac{\text{Volume of air-filled pores}}{\text{Volume of soil}}$$

Water Storage

Within 2 days of a soil being wetted by rain or irrigation, drainage becomes slow and the soil is said to attain its field capacity (*FC*) water content. To measure *FC* accurately, the wet soil surface should be covered to prevent evaporation. The water content at *FC* sets the upper limit for stored water available to the vines and any midrow cover crop. Gradually, as plants extract this water, narrower and narrower pores become air filled until the permanent wilting point (*PWP*) is reached. The amount of water held between *FC* and *PWP*—the soil's "available water"—is an important soil property. Box 4.2 summarizes the different ways of expressing soil water content.

At any given time, the sum of the water-filled and air-filled porosities equals the total porosity. When the soil is at its *FC*, the value of air-filled porosity is called the air capacity, or sometimes the drainable porosity. The volume of available water in a 1-m-deep profile (see box 4.2) is called the available water capacity (*AWC*). These two variables—air capacity and *AWC*—can together be used to classify a soil's structural quality, as shown in figure 4.2.

Box 4.2 Ways of Expressing Soil Water Content

Soil water content is most easily measured by weighing a sample of moist soil, drying it at 105°C for 48 hours in an oven, and reweighing the sample to measure the weight of oven-dry (o.d.) soil. The loss in weight represents the weight of water, expressed as a percentage according to the equation

$$\text{Gravimetric water content (\%)} = \frac{\text{Loss in weight of soil sample}}{\text{Weight of o.d. soil}} \times 100 \quad \text{(B4.2.1)}$$

Another measure of soil water content is the volumetric water content, θ (theta), defined by the equation

$$\theta = \frac{\text{Volume of water-filled pores}}{\text{Volume of soil}} \quad \text{(B4.2.2)}$$

Volumetric water content can be measured directly in the field (see "Monitoring Soil Water"). Also, because 1 Mg water occupies 1 m³ at normal temperatures, a gravimetric water content can be converted to a θ value by multiplying by the soil's bulk density.

Values of θ are useful in the vineyard because they give a direct measure of the "equivalent depth" of water per unit area of the soil. For example, consider a soil volume of 1 m³ that has a θ value of 0.25 m³/m³—equivalent to a water depth of 0.25 m/m² surface area. As shown in figure B4.2.1, we may visualize this as a

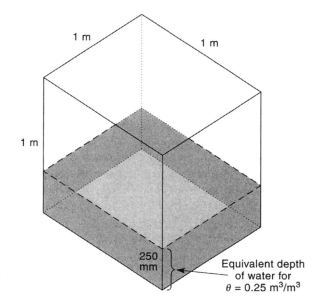

Figure B4.2.1
Equivalent depth of water in a 1 m³ soil volume of water content θ equal to 0.25 m³/m³. (White, 2003)

1 m

1 m

1 m

250 mm

Equivalent depth of water for θ = 0.25 m³/m³

(continued)

Box 4.2 *(continued)*

depth of 250 mm of water in a 1-m depth of soil. Put this way, soil water content is directly comparable with amounts of rainfall, irrigation water, and evaporation, all of which are measured in millimeters.

A θ value of 0.25 m^3/m^3 is the same as 2.5 mm/cm depth of soil. Thus, the equivalent depth d (in millimeters) of water in any soil depth z (in centimeters) is given by

$$d = 10 \times \theta \times z \qquad\qquad\qquad\qquad \text{(B4.2.3)}$$

Table B4.2.1 gives some examples of the use of this equation.

Table B4.2.1 **Some Examples of Equivalent Depths of Water in Soil**

Soil volumetric water content θ (m^3/m^3)	Soil depth z (cm)	Equivalent depth of water d in soil depth z (mm)
0.25	20	50
0.25	40	100
0.40	20	80
0.40	40	160

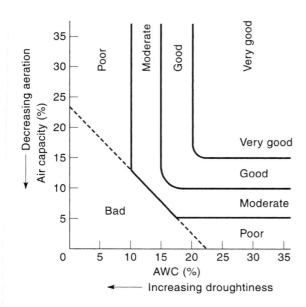

Figure 4.2 A simple classification of structural quality for vineyard soils. (White, 2003)

Soil "droughtiness" increases as the amount of available water decreases, whereas its susceptibility to waterlogging increases as the air capacity decreases. An air capacity of 15% together with an *AWC* of 20% or greater is regarded as being very good. The simple classification of figure 4.2 is useful for vineyard soils because it sets practical limits for good aeration and available water.

Aggregation

The size, shape, and consistence of aggregates vary considerably between soil types and often between the topsoil and subsoil of one soil type. For example, figure 4.3 shows a surface soil in which the degree of aggregation is greater than 90% and the aggregates are mainly blocky but well rounded, and are between 5 mm and 20 mm. This kind of aggregation is typical of loamy soils under grass. The fine roots of the grass plants bind soil particles together, whereas gums and mucilages produced by the roots and associated bacteria and fungi act as a biological adhesive. The more organic matter is in the aggregates, the darker they appear (see, for example, figure 1.10A). This kind of aggregation not only provides good drainage and aeration, but also adequate water storage.

Although grass roots and organic compounds are the key to good surface soil structure, their effect is less significant in the subsoil because they are less abundant there. Good subsoil structure consists of aggregates that are orientated vertically and are longer than broad, with well-defined cracks between and within the aggregates. Ideally, the texture ranges from loam to clay loam to light clay and the color is generally golden brown to red, without mottles, showing that the subsoil is well drained and not waterlogged when wet. The dominant exchangeable cation is Ca^{2+}, with iron (Fe) and aluminum (Al) oxides frequently acting as bridging agents between clay particles. Figure 4.4 shows an example of a well-structured subsoil.

Figure 4.3 Desirable subangular blocky structure of the topsoil of a calcareous soil in Tuscany, Italy. The scale is 10 cm.

Figure 4.4 The well-structured subsoil of a Terra Rossa in the Coonawarra region, South Australia. The scale is 10 cm. (White, 2003)

Forces at Clay Surfaces

The amount and type of clay minerals have an important effect on aggregate formation and stability. As described in "Retention of Nutrients by Clay Minerals and Oxides" in chapter 3, layers within the clay minerals have an overall negative charge that is just balanced by the positive charge of exchangeable cations attracted to the surfaces. The cations available in the soil solution have charges normally ranging from $+1$ to $+3$, and also differ in size according to the number of water molecules in their hydration shells.[a] The cations tend to release these water molecules when attracted close to a clay surface. For example, K^+ has a low hydration energy and sheds its water molecules, Ca^{2+} has an intermediate hydration energy and only partially sheds its water molecules, and Na^+ has a high hydration energy and remains hydrated. Thus, the force between layers and the distance separating layers varies according to the predominant cations present.

Because the cations accumulate in the spaces between clay layers, water molecules try to diffuse into these spaces and, in so doing, they create a swelling pressure that pushes the layers farther apart. This swelling pressure changes with

[a] A hydration shell is an ordered sheath of water molecules surrounding a cation.

the type of cation and the ionic concentration of the soil solution. For example, because three times as many exchangeable Na^+ ions than Al^{3+} ions are required for charge balance in a clay crystal, and the Na^+ ions are highly hydrated, the tendency for water to diffuse into the crystal is much greater in an Na clay than an Al clay. Box 4.3 describes the effect of swelling pressure in more detail.

Box 4.3 What Causes Clay to Disperse?

When Ca^{2+} ions are the main exchangeable cations, clay layers within particles and whole clay particles come close together in parallel alignment because the attractive force between the negatively charged clay surfaces and the cations predominates; the clay is said to be flocculated. However, as Ca^{2+} ions are progressively replaced by Na^+ ions, the weaker negative-to-positive attraction and greater tendency for water to diffuse between the layers causes the particles to swell and move farther apart. Also, the swelling pressure resulting from the influx of water increases if the soil solution becomes more dilute, as happens when the soil is very wet. The net effect is that clay particles separate to a point where the weakened forces of attraction are overwhelmed and the clay deflocculates or disperses.

The clay suspension shown in the middle jar in figure B4.3.1 came from a creek in inland Queensland. Water in many Australian inland streams remains

Figure B4.3.1 A sample of dispersed clay (middle jar) and the same clay flocculated with 0.1 M calcium chloride ($CaCl_2$) solution (right jar). Compare the clear supernatant above the flocculated clay with the clear $CaCl_2$ solution in the left jar.

(continued)

Box 4.3 *(continued)*

"milky" because of the dispersed clay eroded from soils containing exchangeable Na^+. The critical amount of Na^+, expressed as a percentage of the soil's *CEC*, above which dispersion occurs ranges from 6% to 15%, depending on the type of clay mineral present and the soil solution concentration. Dispersed clay is easily transported in runoff water, which worsens the tendency for soil to erode (see "Cover Crops and Soil Water" in this chapter). Furthermore, as the soil surface dries, the dispersed clay forms a hard crust that inhibits water infiltration and seedling emergence.

The jar on the right-hand side of figure B4.3.1 shows the same suspension after a small volume of concentrated $CaCl_2$ solution (from the jar on the left-hand side) was added and mixed. The dispersed clay has flocculated and settled to the bottom, and the water is now clear. Flocculated clay is a prerequisite for the formation of small, stable aggregates, which in turn clump together to form larger aggregates.

The Na–Ca interaction is important for the flocculation of montmorillonite, illite, and vermiculite clays. For kaolinite, flocculation that depends on an attraction between positive charges on the clay edges and negative charges on the flat surfaces is more important. Such flocculation prevails at a pH less than 6 when the edges are positively charged, but breaks down as the pH increases and the edges become negatively charged (see "Retention of Nutrients by Clay Minerals and Oxides," chapter 3).

Aggregate Stability

Apart from an aggregate's size, shape, and color, a key characteristic is its ability to resist "slaking" as it wets up. As water is rapidly absorbed, trapped air exerts a disruptive pressure, augmenting the swelling pressure that develops between layers in the clay particles and around the particles themselves. Altogether, these pressures may exceed the forces holding the aggregate together, causing it to collapse or slake. Because of their particular chemical and biological properties, the aggregates shown in figures 4.3 and 4.4 do not slake.

A further change that may occur when a dry aggregate wets and slakes is that the clay particles disperse and go into suspension (see box 4.3). Figure 4.5 shows a simple test for clay dispersion that can be done in the vineyard.

Soil Strength

Soil consistence refers to aggregate strength, which depends on the forces holding the aggregate together. Consistence changes with water content and is greatest when an aggregate is dry. Aggregate consistence affects the resilience of a soil's structure. The best condition is a firm consistence, between loose and rigid, where

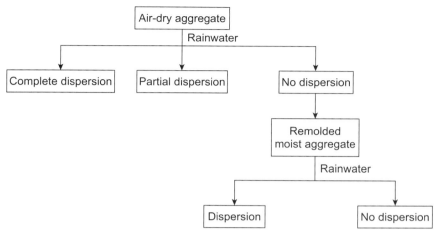

Figure 4.5 A simple dispersion test for soil aggregates, based on the test devised by Emerson (1991), modified by Cass (1999). (Redrawn from White, 2003)

Figure 4.6 Naturally friable tilth in a vineyard soil on the Mornington Peninsula, Victoria, Australia. The scale is 12 cm.

the soil can break down into small aggregates that constitute a friable "tilth." Figure 4.6 shows a sample of undervine soil where surface clods break down easily into firm aggregates 10–30 mm in size.

In addition to the strength of its aggregates, soil has an overall strength that depends on its bulk density and water content. This relationship is shown in figure 4.7, where the units of strength are those of pressure (force per unit area), commonly mega-Pascals (MPa). Box 4.4 describes the measurement of bulk soil strength.

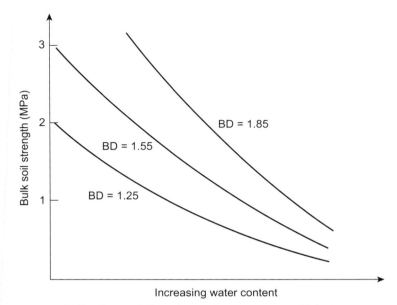

Figure 4.7 Bulk soil strength depends on soil water content and bulk density (BD). (Redrawn from White, 2003)

Ideally, strength in the upper part of a moist soil profile should be between 1 and 2 MPa to provide sufficient load-bearing capacity for machinery, while permitting vine root exploration of the midrow soil. Measured at the *FC*, soil strength should be less than 2 MPa and not exceed 3 MPa at the *PWP*. Permanent grass swards in the mid rows are effective in drying the soil and maintaining soil strength in the optimum range. Winter cover crops have a similar effect during what is usually the wettest period of the year for vineyards.

A soil of inherently low strength is susceptible to compaction if trafficked by machinery when too wet. For machinery traveling between the rows, compaction in the wheel tracks close to the vines can be a problem, but this kind of compaction does not occur with over-the-row machinery because the wheels travel mid row. Where the mid rows are repeatedly cultivated, a compacted layer can develop just below cultivation depth. Generally, compaction is more severe in sandy soils than clay loams and clays, but soils with sodic clay subsoils are an exception to this generalization. In Australia, the critical Na content for a sodic clay, expressed as a percentage of the soil's *CEC*, is ≥6% (see box 4.3).

Examples of problems arising from different forms of soil compaction in vineyards are described in "Cultivation and Ripping" in chapter 2. However, in the case of high-vigor sites, some wheel-track compaction may be of benefit in restricting the access of vine roots to the water and nutrient reserves of the mid rows.

Box 4.4 Measuring Soil Strength in the Vineyard

Soil strength is measured by the resistance offered to penetration with a penetrometer, an instrument consisting of a steel rod about 1 m long with a sharp, conical tip and a pressure gauge at the top (figure B4.4.1). As the tip is pushed into the soil, the pressure applied to overcome the soil's resistance is recorded by the gauge. The rod is graduated so that the pressure can be recorded at intervals down the profile. Readings are most reliable when the rate of entry of the rod is constant. Penetration resistance varies spatially, so a large number of readings is required to obtain a reasonable average. Because the strength depends on soil wetness, measurements should be made at the same water content, usually around *FC*.

Penetrometers are valuable for detecting where compacted soil or hard layers occur, and how severe they are. This information allows the most appropriate ameliorative treatment to be determined, whether it be deep ripping, sowing a permanent cover crop, applying gypsum, mulching, or some combination of these.

Figure B4.4.1 A hand-held penetrometer in use. (Photo courtesy of Dr. Jack Holland, Charles Sturt University, Australia)

Water in Soil

Forces Acting on Water

The two most important forces acting on soil water are gravity and suction. The gravitational force simply depends on the height of the water with respect to some reference level, usually the soil surface. Under the influence of gravity, water lying on the surface will seep into the soil or flow downslope.

The action of the suction force on water is more complex. As discussed previously, an integral part of a soil's structure is the network of connected pores of different sizes and shapes. The narrower a soil pore, the stronger the surface tension force drawing water into that pore. Hence, water in contact with dry soil is preferentially sucked into the narrowest pores, and if the supply is maintained, water progressively fills larger and larger pores until all the pores are filled and the soil is saturated. The last drop of water to enter the largest pore is held at zero suction.

The action of suction and gravity can be illustrated by the behavior of a sponge placed in a shallow dish of water (figure 4.8). Water is sucked into the sponge until it wets to a certain height; this is referred to as capillary rise. The upper part of the sponge does not become wet because the suction drawing water up is canceled out by gravity pulling water down to the "free" water level in the open dish. The net result of these two forces—suction and gravity—determines the hydraulic head of the water. Exactly the same situation occurs in a vineyard; the soil water is at equilibrium (in the absence of surface evaporation) when the suction force on the water is equal but opposite to the gravitational force at each point in the soil profile. The suction force resulting from surface tension at air– water–solid interfaces is called matric suction.

At some level, usually deep in the profile, the soil may be saturated and we find groundwater. The top of this groundwater is called the water table, and the suction force there is zero, as shown in figure 4.9. Relative to atmospheric pressure, water below the water table has a positive hydrostatic pressure that increases with depth. Hydraulic head differences arise because of differences in groundwater

Figure 4.8 Capillary rise of water. The dyed water is less than 1 mm deep in the glass dish and has risen approximately 10 mm into the sponge in a uniform front.

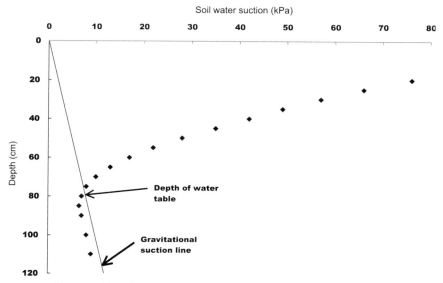

Figure 4.9 The curved line shows how the hydraulic head of water in a soil profile changes with suction and gravitational forces above and below a water table.

depth and land height, with the result that groundwater flows in the direction of minimizing these head differences.

This simple picture of soil water must be modified when there are dissolved salts, as seen in saline soils. Salty soil water has an osmotic suction, with the result that the combination of matric and osmotic suction forces is greater than for pure water in the same situation. Consequently, at the surface of a saline soil, the total force pulling water upward is greater than in a nonsaline soil at the same water content. The osmotic suction effect causes saline groundwater to rise by capillarity from depths of 1 m in sandy soils and up to 2 m in clay loams. Box 4.5 describes the main types of saline (and sodic) soils. The management of soil salinity is discussed later in this chapter in "Drainage, Leaching, and Salinity Control."

Availability of Water to Vines

Soil Infiltration and Wetting

Infiltration starts when water falls on the vineyard soil as rain or irrigation. Both matric suction and gravity pull water into the soil, with suction being dominant initially when the soil is dry, but negligible when it is thoroughly wet. If water is applied rapidly, as with heavy rainfall, the capacity of the soil to accept it may soon

Box 4.5 Types of Saline and Sodic Soils

Soil salinity is conveniently measured by the EC of the soil solution (see "Soil Testing," chapter 3). Most data on soil EC relate to a saturation extract (EC_e), and the distinction between saline and nonsaline soil is set at an EC_e value of 4 dS/m. In Australia, soil EC is measured in a 1:5 soil/water suspension, but these values can be converted to EC_e values using the table in appendix 2.

In a saline soil, the exchangeable cations Ca^{2+} and Mg^{2+} are gradually displaced by cations such as Na^+. The tendency for this to occur is measured by the sodium adsorption ratio (SAR) of the soil solution, defined as

$$SAR = \frac{[Na^+]}{\left[\dfrac{Ca^{2+} + Mg^{2+}}{2} \right]^{1/2}} \qquad (B4.5.1)$$

The ion concentrations in this equation, indicated by the square brackets, are measured in millimoles of charge per liter. When the SAR is ≥ 3, the soil's exchangeable sodium percentage (ESP) is likely to be ≥ 6, and the soil becomes saline–sodic.

Saline–sodic soils are potentially unstable when irrigated because the application of water of low salinity gradually leaches out the resident salts. The lowered salt concentration, coupled with the high ESP, causes the clay to begin dispersing with a consequent destabilization of soil structure. With the removal of salts, the soil becomes simply a sodic soil.

The first sign of structural damage is a decrease in the soil's permeability to water, referred to as its hydraulic conductivity. This effect is most noticeable in a clay subsoil in which dispersed clay can block soil pores, reducing the hydraulic conductivity to a very low value. Waterlogging is likely to ensue.

Sodic soils should be treated with gypsum to replace Na^+ ions with Ca^{2+} and so promote clay flocculation and stable aggregation. Topsoil sodicity is easy to treat with surface applications of gypsum (1–2 t/ha/year) for 2–3 years, but subsoil sodicity may require deep ripping and concurrent gypsum application (see "Cultivation and Ripping," chapter 2).

be exceeded, with the result that water collects on the surface and begins to run off. Running water can cause erosion, especially on steep slopes when the soil is bare. Box 4.6 discusses ways of measuring rainfall.

As infiltration proceeds, a zone of wet soil extends downward from the surface. Figure 4.10 is a good example showing a "wetting front" clearly separating wet soil above from drier soil below. Although suction remains important for drawing water into pores at the wetting front, gravity is the dominant force driving flow through the bulk of the wet zone. Box 4.7 describes how to estimate the depth to which water has penetrated after rain or irrigation.

Box 4.6 How to Measure Rainfall

Rainfall is measured using a rain gauge such as that shown in figure B4.6.1. The gauge should be installed at a standard height (the rim should be 30 cm above the ground in Australia), because wind turbulence causes the amount of rain collected to decrease with the height of the gauge above the ground. The gauge should not be overshadowed by vegetation or buildings, nor too exposed to wind. Read the gauge daily at a standard time, usually 9 AM. Rain is measured in millimeters or inches (for conversion, see appendix 1) and 1 mm of rain is equivalent to 1 L/m² of ground surface.

Rainfall rate or intensity is measured with a recording gauge that usually has two small "buckets," of capacity equivalent to 0.1–0.2 mm of rain, which tip alternately. An electrical switch sends a pulse to a data logger each time a bucket tips, so rainfall intensity can be measured very accurately in millimeters per minute.

Some advisers in Australia and California calculate "effective" rainfall as a fraction (65%–80%) of measured rainfall. The concept of effective rainfall allows for surface runoff, which is obviously rain that does not infiltrate the soil. However, it should not be used to correct for rainfall intercepted by the canopy and subsequently evaporated, because the energy used in evaporating this water is not available to drive transpiration from the leaves. Thus, although the intercepted water does not reach the soil, the vines' transpiration loss is reduced, and the balance between actual *ET* and rainfall does not change.

Growers should use their knowledge of rainfall intensity and vineyard conditions (soil surface condition, mulching, and slope) to decide whether to correct for effective rainfall.

Figure B4.6.1 A standard rain gauge. (White, 2003)

Figure 4.10 A soil profile showing a clear wetting front. (White, 2006; reprinted with permission of Wiley-Blackwell Publishing Ltd.)

Box 4.7 Estimating the Depth of Soil Wetting

Rainfall or irrigation of 20 mm means that water is deposited on the soil surface to a depth of 20 mm. As it infiltrates, this water can only occupy the soil's pore space, which is about 50% by volume (0.5 m^3/m^3). Thus, if all the pore space is initially air filled, the wetting front will penetrate to a depth x_1, where

$$x_1 = 20/0.5 = 40 \text{ mm} \tag{B4.7.1}$$

If there is water in the soil initially (even air-dry soil contains some water), the infiltrating water will penetrate farther. For example, if the initial water content θ is 0.1 m^3/m^3, the effective pore space to be occupied is 0.4 m^3/m^3 and the depth of penetration x_2 is

$$x_2 = 20/0.4 = 50 \text{ mm} \tag{B4.7.2}$$

Some further penetration of water occurs as the wet soil zone drains to its field capacity. Note also that water does not always penetrate as a uniform wetting front because of preferential flow, as discussed under "Soil Infiltration and Wetting" in this chapter.

In clay soils that crack when dry (see figure 4.1), infiltration and wetting do not occur in such an orderly way. Rather, water flows rapidly down the cracks so that the soil columns between the cracks wet up from the sides as well as the top. This type of flow is called preferential flow or bypass flow, implying that the water flows down preferred pathways between aggregates more quickly than within the aggregates.

Another situation causing irregular infiltration is when the surface soil becomes water repellent (figure 4.11). This is most noticeable in sandy soils that have become very dry. Organic matter coating the sand grains creates a hydrophobic film, repelling water that may then "funnel" down old root channels and worm holes. The effect gradually disappears as the soil wets up under persistent rain, and can be ameliorated by mixing clay into the soil.

Redistribution of Water in the Soil

Water continues to flow even after infiltration ceases, but flow becomes progressively slower as the largest pores and cracks empty of water, followed by the next largest and so on. After 2 days in a well-structured soil, the *FC* is reached, as described in "Water Storage" (this chapter). The pores drained of water at *FC* are often called macropores, whereas those that retain water are called micropores.

The matric suction corresponding to *FC* is about 10 kPa. However, vine roots can extract water at suctions up to 1500 kPa, which corresponds to the *PWP*. The choice of 1500 kPa for *PWP* is arbitrary, because wilting depends on

Figure 4.11 Separate water drops remain on the hydrophobic surface of a dry sandy soil (left side). The same soil on the right has been kept moist and accepts water readily.

Table 4.2 **Instruments for Measuring Soil Suction**

Type of instrument	Basic construction	Suction range	Installation and use
Standard tensiometer	Small cylindrical porous cup sealed to a water-filled PVC tube connected to a vacuum gauge (aboveground)	0–80 kPa (or centibars)	Install at different soil depths; reading gives the combined gravity and matric components of soil suction, so must be corrected for depth to give the matric suction.[a] Purge air bubbles before reading.
"Loggable" tensiometer	As for a standard tensiometer except that suction is measured by a pressure transducer connected to a data logger	0–80 kPa	As for a standard tensiometer except that if the pressure transducer is placed next to the porous cup, no correction is needed for gravitational suction.
Gypsum block, Lite or Watermark sensor	Cylinder of porous material and embedded gypsum pellet wrapped in a geotextile sheath; two embedded electrodes to which an alternating current is applied	10–200 kPa	Electrical resistance of the block changes with its water content, which responds to soil suction; the resistance-to-suction relationship is obtained by calibration; affected by salinity greater than 1 dS/m. Correct for depth to obtain matric suction. Most suited to sandy/loamy soils.
Gypsum block, heavy	Solid cylinder of gypsum containing embedded electrodes, as for a Watermark sensor	50–600 kPa	Calibrate as for the Watermark sensor; unaffected by salinity less than 6 dS/m. Correct for depth to obtain matric suction; useful in clay soils.

[a] For a reference level at the soil surface (where the vacuum gauge is located), the matric suction is equal to the tensiometer reading less 1 kPa for every 10 cm of soil depth (see figure 4.9).

a complex interaction between the variety (also whether on own roots or rootstock), the weather, and how fast water can flow through the soil at large suctions. Tensiometers and gypsum blocks are used to measure soil water suction, as explained in table 4.2.

Soil Water Retention and Plant Available Water

Matric suction increases as the diameter of pores holding water decreases, so that grapevines can extract water more easily from wide pores or cracks than from narrow pores. The relationship between the amount of water held and its matric suction is called the soil water retention characteristic, which is an important soil property.

Figure 4.12 gives an example of the water retention curves at matric suctions up to 600 kPa for the top 10 cm and for 40–60 cm of subsoil in a well-structured clay loam. From experience, growers have found that soil water held between 10 and 60 kPa is readily available to the vine (called readily available water, or *RAW*), whereas that held between 60 and 400 kPa is less easily extracted (called deficit available water, or *DAW*).

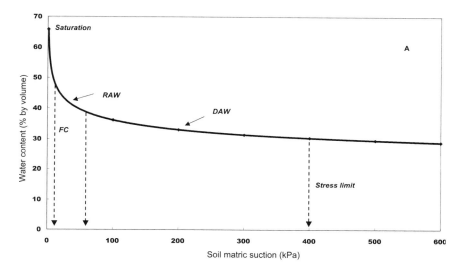

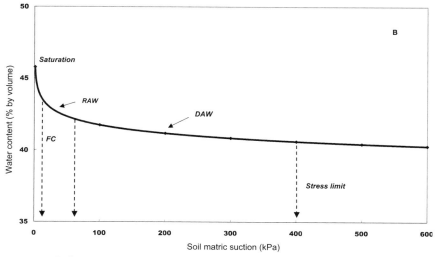

Figure 4.12 Soil water retention curves for a well-structured clay loam. (A) A depth of 0–10 cm in the A horizon. (B) A depth of 40–60 cm in the B horizon. Readily available water (*RAW*) and deficit available water (*DAW*) are shown for both depths. (Original data courtesy of Dr. Lilanga Balachandra, University of Melbourne, Australia)

A suction of 400 kPa is called the stress limit, indicating that although vines can extract water held at greater suctions, the rate of supply is not fast enough for them to function optimally. The stress limit changes with soil texture: it is 100 kPa in sandy soils, 200 kPa in loams/clay loams, and 400 kPa in clay soils. These stress limits apply to mature vines. Vines less than 3 years old should not be subjected to suctions greater than 60 kPa in the midroot zone.

For the top 10 cm of soil in figure 4.12, RAW and DAW were both 9% by volume ($\theta = 0.09$ m³/m³). From equation B4.2.3 in box 4.2 we know that this θ value is equivalent to a depth of water d of

$$d = 10 \times 0.09 \times 10 = 9 \text{ mm}$$

for both RAW and DAW. In the subsoil at 40–60 cm, RAW and DAW were 1.3% and 1.6% ($\theta = 0.013$ and 0.016 m³/m³), respectively, corresponding to 2.6 and 3.2 mm, respectively. Thus we see there is generally more RAW plus DAW in the topsoil than in the subsoil, where the pores are narrower and water is held more tightly.

Values of RAW and DAW should be calculated for the depth of interest in the soil profile, which may be the depth for irrigation management or the maximum rooting depth in dryland vineyards. Table 4.3 shows some calculated results for the clay loam of figure 4.12, assuming that the 0–10-cm data apply to the whole A horizon (40 cm deep) and the 40–60-cm data apply to the whole B horizon (40–120 cm deep). The AWC for the 0–60-cm depth, and the whole profile (0–120 cm), are calculated the same way and give the plant available water (PAW) in these two depths of soil.

Although the relationship between the amount of soil water and its matric suction is mainly influenced by soil structure at small suctions, the effect of texture

Table 4.3 **Calculated Readily Available Water (RAW), Deficit Available Water (DAW), and Plant Available Water (PAW) for the 1.2-m-Deep Profile of the Clay Loam Shown in Figure 4.12**

Soil depth interval (cm)	Soil texture	RAW[a] (mm per depth interval)	DAW[a] (mm per depth interval)	PAW[a] (mm per depth interval)
0–10	Loam	9	9	20
10–40	Clay loam	27	27	60
40–60	Clay	2.6	3.2	12
Totals (in millimeters) for all the available water components to a 60-cm depth		38.6	39.2	92
60–120	Clay	7.8	9.6	36
Totals (in millimeters) for all the available water components in the whole profile		46.4	48.8	128

[a] Calculated using the equation d (mm) $= 10 \times \theta \times z$, where z is measured in centimeters.

Table 4.4 The Range of Readily Available Water (*RAW*), Deficit
Available Water (*DAW*), and Plant Available Water (*PAW*) for Soils of
Different Texture (measured in millimeters per centimeter depth of soil)

Soil texture	*RAW*	*DAW*	*RAW* + *DAW*	*PAW*
Loamy sand	0.55	0.15	0.70	0.86
Sandy loam	0.64	0.28	0.92	1.15
Sandy clay loam	0.71	0.41	1.12	1.43
Clay loam	0.65	0.51	1.16	1.48
Heavy clay	0.41	0.18	0.59	1.20

Compiled from Nicholas (2004).

is more obvious at large suctions. Because texture is a less changeable property than structure, representative values of *RAW*, *DAW*, and *PAW* have been developed for soils of different texture, as shown in table 4.4. This information is used for managing the water supply to grapevines by irrigation, and can also be applied in precision viticulture (see chapter 6).

Managing Soil Water with Irrigation

How Water Stress, Vine Growth, and Grape Flavor Are Related

Water stress is a physiological response to a limitation in water supply. A standard rubric of Old World viticulture is that vines must experience some water stress to enhance the aromas and taste sensations of the wines produced. Better wine quality has been linked with lower yields, so that AOC wines in France, for example, normally cannot be made from irrigated vines. This concept has some following in other countries such as Chile and South Africa, where viticulture has been established for a long time. Nevertheless, irrigation is now used to supplement the water supply to vines in many vineyards around the world where summers are hot and dry, as in central Spain, southern Italy, the Central Valley of California, and the Murray-Murrumbidgee region of inland Australia, because commercial wine production would not be possible without it.

Although there are several ways of assessing vine water stress (described later), here I shall discuss it in terms of soil available water. When soil water is within the *RAW* range, vines are well watered; they have actively growing shoot tips and normal internode expansion. However, if water is too readily available, especially during the period of rapid shoot growth from preflowering to veraison, excess vigor may result, as evidenced by long shoots and a large leaf area and density that cause excessive fruit shading (figure 4.13). Vines with excess vigor are "out of balance" with respect to the amount of vegetative growth relative to fruit carried.

When moderate water stress occurs (soil water content at the bottom of the *DAW* range), shoot growth slows, internodes shorten, and shoot tips may become

Figure 4.13 Ten-year-old Shiraz vines showing vigorous growth under subsurface drip irrigation in McLaren Vale, South Australia.

Figure 4.14 Cabernet Sauvignon vines showing signs of severe water stress during late summer in Napa Valley, California, USA.

a dull gray-green. At the hottest part of the day, leaves become flaccid (temporary wilting) and warm to touch. With more severe stress (soil water below the *DAW* range), shoot growth stops and the tips and tendrils may die. Late in the season, water-stressed basal leaves become yellow and may develop necrotic areas toward the edges. Early senescence and leaf fall begin at the base of a shoot and progress toward the tip. Figure 4.14 shows an example of water-stressed vines in late summer in Napa Valley, California.

Water stress at flowering can reduce fruit set. From fruit set to veraison, moderate to severe stress reduces berry size through its effect on cell division and cell enlargement. During ripening, mild water stress enhances the accumulation of soluble solids in the berries by suppressing vegetative growth, but more severe stress decreases berry size through its effect on cell expansion (commonly seen as "berry shrivel"). In this case, sugar accumulation and flavor development are delayed as a result of decreased photosynthesis and premature leaf fall.

In cool humid climates such as Burgundy, and humid maritime climates such as the Médoc, dryland vines do not usually experience water stress from flowering to veraison because the supply of soil water is adequate. During the final ripening period, however, a gradual increase in stress enhances the intensity of flavors in the berries, so that outstanding vintages are produced in years of dry and hot mid to late summers. This natural response to a changing water supply seldom occurs in hot inland regions where irrigation is essential to grow grapes. Instead, growers rely on manipulating the amount and timing of irrigation to control vigor and enhance grape quality, as discussed later.

How a Soil Water Deficit Develops

■   Evaporation and Transpiration

In winter-rainfall Mediterranean and cool, humid regions, the soil profile under dormant vines is normally at FC by the end of winter. During spring, evaporation rates increase as a result of more sunshine and rising temperatures. Evaporation is driven by radiant energy from the sun.

A fraction of the solar radiation received is directly reflected, depending on a surface's reflectance or albedo (literally, its "whiteness"). Some of the short-wave radiation absorbed by the soil and vegetation is reradiated to the atmosphere as long-wave radiation. The greatest part of the remaining "net radiation" is dissipated through the evaporation of water, and the rest is partitioned between heat transmitted to the air aboveground and heat penetrating deeper into the soil. Heat transferred to the air aids grape ripening whereas heat transferred into the ground raises the soil temperature. Figure 4.15 shows a summary of these energy transfers.

Water evaporates from bare moist soil at a maximum rate determined by the evaporative demand of the atmosphere until the surface 1–2 cm begins to dry. Similarly, after bud burst occurs and leaves expand, the vine transpires water at a rate determined by the air's evaporative demand. The combined effect of these "atmosphere-driven" processes produces the potential evapotranspiration (PET), the rate of which can be estimated as described in box 4.8. As the surface layer becomes air-dry, the soil evaporation rate slows virtually to zero.

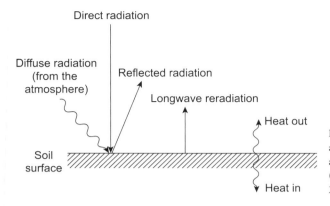

Figure 4.15 Radiation and heat energy balance at a bare soil surface. (Redrawn from White, 2003)

Box 4.8 Estimating Evapotranspiration

The *PET* of a vegetated soil surface depends on the radiant energy absorbed, the relative humidity of the air, and the wind speed. These are called meteorological variables, the values of which are measured with a weather station (figure B4.8.1). The calculation of *PET* assumes a reference crop of green grass 12 cm high and completely covering the soil surface, with an unlimited supply of water. Data for *PET* (measured in millimeters per day) are supplied by meteorological services

Figure B4.8.1 Example of a weather station in a vineyard. (White, 2003)

(continued)

Box 4.8 *(continued)*

(e.g., in Australia, the SILO data drill of the Bureau of Meteorology [www.bom. gov.au/silo]).

A simple way to measure evaporation is with a shallow tank of water exposed to wind and sun, such as the widely used class A pan shown in figure B4.8.2. The pan is raised above the soil surface to allow air to circulate and so minimize overheating. The daily fall in water level (after allowing for any rain) is measured to give the pan evaporation rate E_p (measured in millimeters per day). Pan evaporation rates are usually about 20% greater than PET at the same site.

Even when there is an adequate supply of soil water, ET from a vineyard does not necessarily occur at the potential rate because the canopy may not be fully developed. Research in California has shown that water use by well-watered vines is linearly related to the percentage of the soil surface shaded by the canopy (at mid-day), up to a maximum of 80%. To account for this effect, PET is adjusted with a crop coefficient C_c to give an adjusted ET as follows

$$\text{Adjusted } ET = PET \times C_c \tag{B4.8.1}$$

Figure B4.8.2 A class A evaporation pan. The pan diameter is 1.2 m and the depth is 254 mm. The water level is kept 54 mm from the top rim.

(continued)

Box 4.8 *(continued)*

Similarly, if E_p data only are available, the adjusted ET is obtained using a crop factor C_f as follows

Adjusted $ET = E_p \times C_f$ (B4.8.2)

Table B4.8.1 gives an example of crop coefficients and crop factors for two vineyards in southern Australia. Note that if RDI is implemented, the coefficients or factors are scaled down during RDI because when some water stress occurs, the actual ET is less than the adjusted ET.

Crop factors and coefficients may need to be adjusted for local conditions, depending on site potential, vine spacing, the presence of a cover crop, and mulching. For relatively small blocks in hot, dry regions, the crop coefficient for a fully irrigated vineyard at full canopy may be 1.1–1.2 because of the extra energy transferred to the canopy by hot winds blowing from the dry surroundings.

Table B4.8.1 Crop Factors and Crop Coefficients for Vines at Different Growth Stages in a Temperate Climate

At the stage of...	Crop factor C_f^a		Crop coefficient C_c^b	
	No stress	RDI[c]	No stress	RDI[c]
Bud burst	0.15	0.15	0.2	0.2
Flowering	0.35	0.35	0.45	0.45
Fruit set	0.45	0.25	0.6	0.3
Veraison	0.50	0.25	0.6	0.3
Harvest	0.50	0.5	0.6	0.6
Postharvest	0.25	0.25	0.3	0.3

[a] The crop factors are for healthy vines growing on a sandy loam soil on a single trellis up to 2 m high, with a mown, midrow cover crop. The canopy coverage increases from 10% of the ground at bud burst to 40% at veraison.

[b] The crop coefficients are for similarly managed vines and canopy cover, except that the mid rows are bare. Crop coefficients for vines with a cover crop would be slightly larger.

[c] Regulated deficit irrigation (RDI) applied from fruit set to veraison to maintain soil water supply between the no stress and stress limit (see "Soil Water Retention and Plant-Available Water," this chapter).

Compiled from Nicholas (2004), Prichard and Verdegaal (2001), and Goodwin (1995).

A soil water deficit (*SWD*) develops as water is withdrawn by evapotranspiration (*ET*). A vine accessing *RAW* normally continues to transpire at or near the maximum rate, and the deeper the roots have penetrated the more *RAW* can be accessed. However, as this source is depleted, the vine begins to draw on *DAW* and to experience some stress; the leaf stomata start to close. The greater the stress, the longer the stomata stay shut, thus limiting photosynthesis and sugar production. During this latter phase, the actual *ET* is considerably less than the *PET*.

Without intervention, the seasonal trend in *SWD* in vineyards under winter rainfall is similar to a sine wave—that is, the deficit increases during summer and then decreases as winter rains refill the soil profile, only for the cycle to be repeated the following season. Figure 4.16 shows an example of this seasonal fluctuation in *SWD*. Note that the size of the *SWD* depends on several factors:

- Soil type and the effective rooting depth of the vines
- Leaf area per hectare (related to the number of vines per hectare and their management)
- Whether mulch and/or cover crops are used
- The balance between rainfall and actual *ET* during the growing period

Controlling *SWD* within limits can bring benefits through decreased vine vigor and, for red varieties in particular, improved fruit quality. When *SWD* is controlled, there is less need for summer trimming and leaf removal. In addition, control of the amount of irrigation water applied not only conserves this increasingly valuable commodity, but also reduces pumping costs.

Deficit Irrigation

Regulated deficit irrigation (RDI) is the most common way of scheduling irrigation to supply less than a vine's full water needs during part of the growing season. Regulated deficit irrigation enables growers to determine when and how much

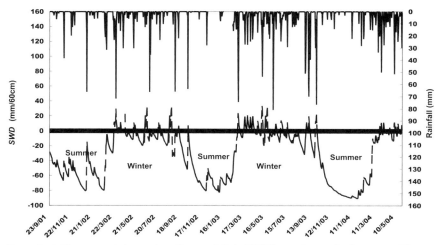

Figure 4.16 The seasonal change in soil water deficit (*SWD*) over 3 years for Sauvignon Blanc vines without irrigation in a cool-climate vineyard in Victoria. Rainfall, solid lines; *SWD*, dashed line. When the *SWD* is positive, drainage can occur. (Original data courtesy of Dr. Lilanga Balachandra, University of Melbourne, Australia)

water to apply to meet their objectives for crop yield and fruit quality. Within these broad objectives, there may be specific objectives depending on the variety (and the scion–rootstock combination) and the style of wine to be made. For example, Shiraz is a vigorous variety often requiring a more severe RDI regime to optimize quality than Merlot, which is more sensitive to water stress. Cabernet Sauvignon is intermediate. Furthermore, rootstocks confer differential tolerance to water stress on the scion varieties, as discussed in "Rootstocks" in chapter 5.

The preferred RDI method uses a threshold deficit, which is the deficit that, when exceeded, signals the need for irrigation. In Australia, RDI is usually applied between fruit set and veraison (see box 3.5), when the threshold deficit for irrigating is equated to *RAW* plus some fraction of *DAW*. Normally no more than the top 60 cm of soil is deficit regulated. For the example given in table 4.3, the threshold could be set at *RAW* plus one half of *DAW*, giving a total of 58 mm, which is approximately 60% of *PAW* to a 60-cm depth. A smaller deficit is advocated from veraison to harvest so that shoot growth is controlled, but sugar accumulation in the fruit is not inhibited. Methods of applying the water are discussed later in "Irrigation Methods."

Sustained deficit irrigation is another aspect of deficit irrigation that is attracting more attention in the inland regions of southeastern Australia, where irrigation water is in short supply. This approach is analogous to the "volume balance" concept, as practiced in California's Central Valley. There, the threshold deficit is set by the leaf water potential (discussed later) and, after the threshold is reached, 60%–70% of the vine's full demand is supplied each week until after harvest. The vines' full demand is assessed from *PET* data.

Methods of monitoring vine water stress through soil or plant measurements are discussed in the following paragraphs.

Monitoring Soil Water

One way of monitoring soil water is to measure matric suction with tensiometers and/or gypsum blocks, as explained in table 4.2. For suction measurements, the trigger for irrigation to start is set between 60 and 400 kPa, depending on the soil's texture and the degree of stress to be imposed.

Another way is to measure the volumetric water content θ. For θ measurements, the preferred instruments are those that allow repeated nondestructive sampling, such as capacitance probes or time domain reflectometry (TDR). Neutron probes are generally not used now for health and safety reasons. Box 4.9 describes this equipment in more detail.

With all these instruments, it is important to realize that the values obtained depend on soil type, depth, and distance from the vine. Because all vineyards show some soil variation, there should be at least one measuring site per soil type with at least two depths of measurement—one in the zone of maximum root volume at a 20–30-cm depth, and the other near the base of the "managed" root zone

---

Box 4.9 Instruments for Measuring Soil Water Content

In dryland vineyards, soil water content should be measured at two to three depths, down to the bottom of the root zone. In irrigated vineyards, the recommended practice is to manage the water content within the top 60 cm of soil, so this defines the depth of interest for monitoring.

The capacitance probe and TDR rely on measuring the change in the apparent dielectric constant of soil with a change in water content (water has a very high dielectric constant compared with soil solids and air). A capacitance probe can reside permanently in a narrow-bore PVC tube installed in soil, or can be used as a portable instrument lowered to different depths in a PVC tube. Correct installation of the tube is important to avoid air gaps between the tube and surrounding soil; otherwise, false readings are obtained. The annulus of soil that is sensed around the tube wall is small.

For TDR, an electromagnetic pulse is transmitted down parallel steel wave-guides, up to 30 cm long, inserted into the soil. The time taken for the reflected pulse to return to a receiver is proportional to the soil's apparent dielectric constant, and hence varies with the water content. Time domain reflectometry probes may be installed permanently or may be portable if the pulse generator and receiver are built into the probe head. The TDR measures an average θ in a narrow cylinder of soil surrounding the wave guides.

Nicholas (2004), Prichard and others (2004), and Charlesworth (2000) give further details of these instruments.

(at 50 cm). More sites may be needed in undulating land where soil water content is likely to vary between the top and bottom of slopes, but a compromise must be made between the cost of more sensors and the value of the extra information obtained. Tensiometers, gypsum blocks, or access tubes for a capacitance probe should be placed in the rows and between vines at a distance of 10–15 cm (sandy soil) or 20–25 cm (clay soil) from an emitter or mini sprinkler.

Box 4.10 describes the calculation of an *SWD* from measurements of soil water content or *ET*, and its use in irrigation scheduling. Implementing an RDI program to take account of soil variability is discussed in "Precision Viticulture" in chapter 6.

■ Plant Measurements of Water Status

A common plant-based measurement of water status focuses on leaf or stem turgor, which reflects the suction force on water in the vine. This suction force is expressed as a water potential, a negative variable, which is measured as either a stem water potential (*SWP*) or a leaf water potential (*LWP*). For *LWP*, a young mature leaf in the upper canopy is enclosed briefly in a polythene bag before being plucked and placed in a pressure chamber with the cut end of the petiole

Box 4.10 Use of a Soil Water Deficit to Implement Irrigation Scheduling

The *SWD* is calculated from soil water contents or *ET* data. In both cases, assume the soil regains *FC* (the "full point") by bud burst as a result of winter rain or irrigation. The water content (in millimeters) in the full profile at *FC* is best calculated from the average of θ measurements toward the end of winter (use equation B4.2.3 in box 4.2), or it can be estimated from the textures of soil layers as indicated in table 4.4.

If soil monitoring is used, the profile water content is measured at regular intervals during spring through summer (automated systems may provide water content readings on an hourly basis). The *SWD* is calculated as the difference between *FC* and the profile water content on each occasion. Figure 4.16 showed the seasonal trend in *SWD*, calculated in this way, in a non-irrigated block of vines. When the accumulated deficit reaches the threshold value, which might be equal to *RAW* plus half *DAW*, the trigger point for irrigation is reached. Figure B4.10.1 shows the effect of RDI, applied from fruit set to veraison, on the *SWD* developed in the same vineyard soil as shown in figure 4.16.

If *ET* data are used, *FC* at bud burst must be measured or estimated as noted earlier, and the subsequent development of an *SWD* tracked using weekly rainfall

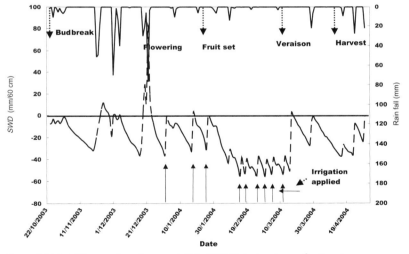

Figure B4.10.1 Changes in soil water deficit (*SWD*) during 1 year for Sauvignon Blanc vines under regulated deficit irrigation in a cool-climate vineyard. With irrigation, the *SWD* was kept in the readily available water (*RAW*) range (0–39 mm/60-cm depth) until fruit set, then in the deficit available water (*DAW*) range (39–58 mm/60-cm depth) up to veraison. The total water applied was 298 mm. Rainfall, solid line; *SWD,* dashed line. (Original data courtesy of Dr. Lilanga Balachandra, University of Melbourne, Australia)

(continued)

Box 4.10 *(continued)*

and *ET* data. For example, if the effective rainfall in the first week of bud burst is P_1 millimeters and the actual *ET* is ET_1, the *SWD* at week's end is given by

$$SWD_1 = P_1 - ET_1 \qquad\qquad (B4.10.1)$$

Rainfall is measured and actual *ET* is estimated as described in boxes 4.6 and 4.8, respectively. Note that if $P_1 > ET_1$, there is surplus water that can be lost as drainage or runoff; conversely, if $P_1 < ET_1$, a deficit develops. For the second week of bud burst when rainfall is P_2 and *ET* is ET_2, the *SWD* is calculated as

$$SWD_2 = SWD_1 + (P_2 - ET_2) \qquad\qquad (B4.10.2)$$

Appendix 4 shows how this calculation can be repeated for subsequent weeks and the results accumulated through to harvest. As shown in figure B4.10.1, the trigger point for irrigation occurs when the threshold *SWD* is reached.

The *ET* method is widely used in California's Central Valley, where the growing season weather is very reliable. Historical *ET* data are entered into the *SWD* equation at the start of each week, and irrigation is scheduled accordingly. Real-time *PET* data, gathered from 120 weather stations distributed around the state, is supplied at the end of the week by the California Irrigation Management Information System (www.cimis.water.ca.gov). Growers adjust *PET* to actual *ET* so that the *SWD* can be recalculated and a correction made the following week for any shortfall or excess of irrigation water applied previously.

A similar system of automated weather stations on a smaller scale now exists in the South Australian region of the Murray River to supply *PET* data to irrigators. Daily values of *PET*, calculated from meteorological observations, can be viewed at www.samdbnrm.sa.gov.au/Portals/7/AWMN/awsview.php.

---

emerging (figure 4.17). The gas pressure just sufficient to force sap from the cut end is assumed to be equal but opposite to the *LWP*. For *SWP*, a leaf near the trunk is enclosed for at least an hour to prevent transpiration and to allow it to equilibrate with the rest of the vine. It is then plucked and measured as for *LWP*. Stem water potential has the advantage of being essentially independent of the leaf's position on a main branch.

Other techniques are being developed to measure vine water status directly, such as a dendrometer to continuously measure small changes in trunk diameter, enabling a grower to make prompt decisions about when to irrigate. Changes in the rate of trunk growth and diurnal changes in trunk diameter correlate well with changes in *LWP*. Another technique uses infrared thermography to measure canopy temperature in the vineyard, which increases as the vine transpiration rate increases. However, only *LWP* is used routinely in commercial vineyards, such as in California's Central Valley. Research there has shown that shoot growth starts to slow when the midday *LWP* falls to –0.6 MPa or less, but photosynthesis is

Figure 4.17 Pressure chamber and accompanying gas cylinder for measuring leaf water potential. (White, 2003)

not affected until −1 MPa is reached. Vines in the range of −1 to −1.2 MPa are only mildly stressed. The normal threshold to trigger RDI is an $LWP \leq -1.2$ MPa, which is approximately equivalent to supplying 60% of full water demand in the Californian "volume balance" approach.

■ Comparison of Methods

Table 4.5 provides a summary of the advantages and disadvantages of soil, weather, and plant-based measurements for assessing vine water status and scheduling irrigation.

Partial Root Zone Drying

Partial root-zone drying (PRD) is a technique that shows promise for regulating vine growth and saving irrigation water without sacrificing grape yield and quality. The basic principle of PRD is to keep part of a vine's root system moist while the remainder is allowed to dry out. A hormone induced in the roots in the dry soil "tricks" the vine into responding as if it were suffering water stress by partial closure of its stomata, which in turn decreases transpiration and slows shoot growth. To sustain the regulating effect, the wet and dry zones need to be alternated; otherwise, the vine adjusts by limiting root growth on the dry side and expanding it on the wet side. Alternation is achieved by having an emitter with its own water supply on each side of individual vines in a row.

Table 4.5 **Methods of Assessing Vine Water Status**

Method	Advantages	Disadvantages
Weather based		
Evaporation pan	Simple and direct daily measurement; applies over a whole vineyard	Indirectly related to vine water status; not variety specific Birds may interfere; crop factor needs to be known for local conditions
Potential evapotranspiration	Can be automated; applies over a whole vineyard	Requires a power source; crop coefficient needs to be known for local conditions
Soil based		
Tensiometers	Can be variety specific Most sensitive in the *RAW* range; pressure transducers are "loggable"; minimal soil disturbance on installation	Indirectly related to vine water status; site-specific measurement Not suitable in the *DAW* range; water may freeze; require a soil water retention curve to calculate how much water is required
Gypsum blocks	Suitable for full range of regulated deficit irrigation; cheap and loggable	Installation disturbs the soil; require a soil water retention curve to calculate how much water is required
Capacitance probe	Measures the full range of θ at all depths; loggable (fixed type) or portable; minimal soil disturbance on installation	Air gaps around access tube affect the reading; fixed type is expensive; readings with portable probe are time-consuming
Time domain reflectometry	As for a capacitance probe	Installation disturbs the soil; readings affected by high salinity
Plant based		
Leaf water potential (*LWP*)	Direct measurement of vine water status	Variable from leaf to leaf and vine to vine; time-of-day dependent; cannot be automated; site-specific measurement
Stem water potential	Direct measurement of vine water status; more sensitive to water deficit than *LWP*	Cannot be automated; time-of-day dependent; labor-intensive
Infrared thermography	Directly correlated with leaf temperature; non-destructive and can be automated	Correlation with leaf temperature is specific to soil, weather, and variety; requires complex calibration

Compiled from Shackel (2006), Charlesworth (2000), and Goodwin (1995).

Early PRD systems had two drip lines in parallel, but an easier to manage system now consists of two lines molded together yet working independently. The alternating cycle of watering needs to be flexible, according to the soil and seasonal conditions. Free-draining soils give the best results because the nonirrigated soil dries quickly and the wet and dry zones are easy to keep separate. In hot, dry weather, the cycle may be as short as 3–5 days, whereas in mild weather it could be as long as 14 days. Soil water sensors should be installed to monitor the rate of soil drying.

In Australia, PRD works best in the hot inland regions where vines rely on irrigation for most of their water during the growing season. The main benefit of PRD is a water saving of up to 50%. Generally, yields are maintained or slightly decreased whereas berry sugar content and titratable acidity are improved. There may also be benefits for soil structure because the wetting and drying cycle can enhance stable aggregate formation.

The Interaction among Cover Crops, Mulches, and Soil Water

Cover Crops and Soil Water

Cover crops grown in the mid rows have a significant effect on the soil water supply to vines. A winter cover crop takes up soil water and, in drying the soil, improves its strength and trafficability (see "Soil Strength" earlier in this chapter). Figure 4.18 shows a cover crop of winter cereal in the western Cape Province region, South Africa. In drying the soil, a cover crop helps to control excess vigor in vines early in the season, which is important on high-potential sites (see chapter 2). Permanent grass cover crops decrease the amount of mineral N available to the vines and lessen the need for herbicides to control weeds.

Figure 4.18 A vigorous cover crop of winter cereal near Stellenbosch, Cape Province, South Africa.

In dryland vineyards, or those receiving minimal irrigation, permanent cover crops are not recommended because soil water needs to be conserved during summer. For this reason, winter cover crops, which in spring are mown or cultivated into the soil, are favored in many vineyards. Keeping the mid rows well mown or free of vegetation in spring–early summer reduces ET and the risk of late frost damage to the vines.

Mulches, Soil Water, and Temperature

When a cover crop is mown in spring, the mowings can be thrown sideways to form a mulch under the vines. Other forms of mulch include compost, bark chips, and cereal straw (figure 4.19). About 50 m³ of mulch is required for an undervine strip 25 cm wide and 5 cm deep at 2000 vines per hectare.

Mulches have several beneficial effects, including:

- Shading the soil, thereby reducing soil evaporation
- Suppressing weeds
- Preventing raindrops from impacting directly on the soil surface, thereby maintaining soil structure and water infiltration
- Encouraging biological activity, especially of earthworms (see chapter 5), through the added organic matter and by keeping the soil moist
- Supplying nutrients that should be accounted for in the fertilizer program

The effect of mulch on soil evaporation is more important when the vines are dormant, and before the canopy is fully developed in spring–early summer. When the proportion of ground shaded by the canopy is less than 10%, more than 70% of the total ET is soil evaporation. However, at full canopy, when the coverage is 50% or more, soil evaporation composes only a small fraction of the total ET.

Figure 4.19 A straw mulch in vine rows in the Barossa Valley, South Australia. (White, 2003)

Mulches affect soil temperature. Because cover crops and mulches have a higher albedo than bare soil, the energy absorbed during the day, and hence soil warming, is less under these types of cover. The fact that soil stays more moist under mulch also influences its temperature, because a wet soil requires more heat energy than dry soil to raise its temperature by a given degree. Soil wetness is therefore important in determining when root growth starts in spring. Vine roots start to grow around 6°C and have a temperature optimum close to 30°C.

Surface albedo is also affected by color, with black and dark-red surfaces reflecting less radiation and therefore warming more quickly than light-colored surfaces. But at night, dark surfaces radiate energy faster than light-colored surfaces, adding to the loss of heat by convection. Bare soil surfaces with no vegetative cover lose heat rapidly at night, especially if there is no low cloud cover. However, the vine canopy traps heat emanating from the soil after dark and reduces heat loss, so that the surface soil temperature and near-surface air temperature remain higher than otherwise expected. This is an important factor in decreasing frost damage to green tissues.

Organic mulches such as straw or bark pieces have a low thermal conductivity and therefore slow the transfer of heat in to and out of the soil. Soil under mulch therefore remains cooler during the day and warmer at night than bare soil. Lower daytime soil temperatures decrease the evaporation rate and favor earthworm activity.

Drainage, Leaching, and Salinity Control

What Causes Poor Drainage?

The best vineyard soils are invariably free draining. They are often deep (e.g., see figure 1.2), but shallow soils on limestone can also be free draining, provided the limestone is porous (as in the Coonawarra region, South Australia) or fractured and fissured (as in the Côte d'Or and St. Emilion regions, France).

An impermeable subsurface horizon, which is often a poorly structured and dense B horizon, impedes drainage. The sodic clay subsoils of some duplex soils in southern Australia are an example, as are some of the duplex soils in the Western Cape Province, South Africa, and the Piedmont region of the eastern United States. In winter, when rainfall exceeds ET for lengthy periods, water accumulates at the top of the B horizon to form a "perched" water table, as was shown in figure 1.1. The soil above the B horizon becomes waterlogged.

Because the solubility of O_2 in water is low, O_2 diffusion through the water-filled pore space is very slow compared with its diffusion in air. As a result, roots and soil microorganisms are starved of O_2 and cannot respire normally. Although some microorganisms, such as those responsible for denitrification (box 4.11), can switch from aerobic to anaerobic respiration and continue to function, vine

Box 4.11   Chemical Reactions in Waterlogged Soil

Microorganisms that respire anaerobically obtain energy for growth by fermenting carbohydrates. The products of respiration are organic acids, such as acetic acid, and the gases carbon dioxide (CO_2) and methane (CH_4). Instead of being transferred to oxygen (O_2) (which is not available), electrons produced during fermentation are transferred to other ions and compounds (called acceptors) that are then reduced.

After O_2, nitrate (NO_3^-) is the most easily reduced acceptor. In fact, NO_3^- reduction or denitrification can occur in partially waterlogged soil, when O_2 is still present in the large pores between the aggregates but is absent from the narrow, water-filled pores inside aggregates. Nitrate is converted to nitrous oxide (N_2O), a potent "greenhouse gas," and dinitrogen (N_2) gas in proportions that vary according to the soil pH, temperature, and intensity of the reducing conditions.

As NO_3^- is depleted, reducing conditions intensify and other compounds are reduced in the following sequence—manganese dioxide (MnO_2), hydrated iron oxides ($Fe(OH)_3$), and finally sulfate (SO_4^{2-}) ions. Reduction of insoluble MnO_2 and $Fe(OH)_3$ produces more soluble Mn^{2+} and Fe^{2+} compounds, whereas SO_4^{2-} reduction produces the "rotten egg" gas (H_2S) typical of organic-rich sediments that have experienced prolonged waterlogging. The reduced S^{2-} ions also combine with the Fe^{2+} ions present to form ferrous sulfide, which precipitates and slowly reverts to iron pyrite (FeS_2), a mineral commonly found in organic-rich river and estuarine sediments and some sedimentary rocks. Exposure to air through disturbance or drainage causes FeS_2 to oxidize and produce Fe^{3+} ions and sulfuric acid (H_2SO_4), which is the basis for the formation of an "acid sulfate soil."

Color changes associated with the reduction and subsequent reoxidation of Mn and Fe oxides are good indicators of whether a soil is permanently waterlogged or simply experiences a fluctuating water table (see table 1.2).

roots cannot function under such conditions, and nutrient uptake and growth of the whole plant suffer.

Although well-drained soil should not develop a perched water table, the rise of regional groundwater to within 1–2 m of the soil surface (see "Forces Acting on the Water" earlier this chapter) can create problems. Because water is pushing up from deep in the profile where there is little organic matter for microorganisms to feed on, the reduction reactions described in box 4.11 are less problematic. However, too much water supplied to the deeper roots encourages excess vigor. Figure 4.20 shows a deep, sandy soil at the bottom of a slope in St. Emilion where this problem occurs. This condition also occurs in some of the heavier textured soils of Pomerol in the Bordeaux region. Furthermore, if the groundwater is saline or salts have accumulated in the soil as a result of poor irrigation management, the capillary rise of salts can have a serious effect on vine health, crop yield, and fruit quality.

Figure 4.20 A deep, sandy soil under vines at St. Emilion, Bordeaux region, France. Note the deep rooting and moist soil in the lower profile.

How to Improve Soil Drainage

Poor drainage resulting in waterlogging can be caused by soil compaction and poor structure. Chapter 2 discusses deep ripping to break up compacted soil when a vineyard is established, and methods of alleviating compaction in the mid rows of established vineyards. Apart from mechanical intervention, soil structure and hence drainage can be improved by applying amendments such as gypsum or by growing a cover crop. Grasses and cereals are most effective because of their fibrous root systems, but the deep taproots of species such as chicory are also effective. Mulches also improve structure through the addition of organic matter and stimulation of soil biological activity.

Mounding soil undervine increases topsoil depth, thereby potentially increasing the rooting volume. Mounds also help to shed surface water more rapidly, especially on slopes, as shown in figure 4.21. Subsurface drainage may be needed to control waterlogging in flat land. In long-established regions, porous clay pipes (called tiles) were laid in rows at a depth between 1 and 2 m to remove excess water and control the water table. For new vineyards, perforated PVC pipe ("aggie" pipe) can be laid by a laser-guided machine on a precise gradient without trenches being excavated. The pipe spacing varies between 10 and 40 m, depending on the

Figure 4.21 Steep mounding of a granite-derived soil in a new vineyard near Stellenbosch, Cape Province, South Africa.

depth at which the pipes are laid and the soil's permeability. Subsoil drainage with pipes has been used, for example, in Pomerol, Bordeaux region, in Tuscany, and in the Carneros region of Napa Valley, California. Although subsoil drainage is expensive, it is often worth the investment, considering the commercial life of a vineyard.

Subsoil drainage is often necessary in irrigated vineyards of the Griffith region, New South Wales, to prevent saline groundwater rising into the root zone. In some cases, groundwater pumping and discharge into evaporation basins may also be required.

Leaching and Salinity Control

Irrigation supplies both water and salts to vines. However, because most of the applied water is transpired and only some of the salts are absorbed, the salt concentration in the soil gradually increases as it passes through repeated wetting and drying cycles. This salt increase can be monitored by measuring EC at the bottom of the root zone, preferably by obtaining a sample of the soil solution with a suction sampler, such as the SoluSAMPLER (figure 4.22). Significant salt accumulation occurs even with low EC irrigation water (<0.75 dS/m), but the problem is worse when water of higher salinity is used. Examples occur with some of the groundwater used in the Barossa Valley, South Australia, with Darling River water in New South Wales, and with Taco River water in Castilla la Mancha, Spain. Sometimes a white salt encrustation can be seen on the soil surface, as shown in figure 4.23.

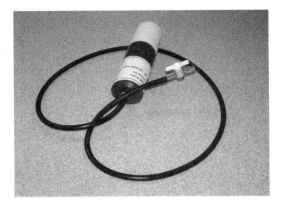

Figure 4.22 The SoluSAMPLER is a porous ceramic cylinder sealed to PVC tubing with a vacuum tap inline. Suctions as much as 60–70 kPa are applied with a syringe attached to the open tube. (Photo courtesy of Dr. Tapas Biswas, South Australian Research and Development Institute, Adelaide, South Australia)

Figure 4.23 A white salt encrustation under a dripper line in a vineyard in Castilla la Mancha, Spain.

Salt accumulation is counteracted by controlled leaching to maintain a salt balance, using a leaching requirement (LR), as discussed in box 4.12. The LR can be met through regular irrigation scheduling. However, with watering strategies such as RDI or PRD under drip irrigation, deep drainage during the irrigation periods should be minimal. Leaching of accumulated salts then needs to occur during the winter dormant period, either through rainfall or a deliberate blanket irrigation.

It is important to distinguish between LR and leaching fraction (LF), the latter being the actual fraction of applied water that passes through the root zone. Ideally LF should equal LR, but because of the variability in water application and infiltration, LF is usually greater than LR, which means more water drains through the root zone in some parts of the vineyard than is necessary for salinity control. This occurs particularly with flood and furrow irrigation (see "Irrigation Methods" later in the chapter).

Box 4.12 Salinity Control by Leaching

To maintain a salt balance in the soil, the input and output of salts must be equal. For an irrigated vineyard, accession of salts in rainfall and from fertilizer is approximately canceled out by the salts removed in fruit (and possibly prunings) and precipitation of insoluble compounds, so the salt balance simplifies to

$$\text{Salts from irrigation water} = \text{Salts in drainage below the root zone} \qquad \text{(B4.12.1)}$$

Because EC is a surrogate measure of salt concentration, and volume per unit area is equivalent to a depth d of water (in millimeters), the salt balance can be written as

$$EC_{iw}d_{iw} = EC_{dw}d_{dw} \qquad \text{(B4.12.2)}$$

where the subscripts iw and dw refer to irrigation and drainage water, respectively. By rearranging this equation, the leaching requirement (LR) can be equated to a ratio of EC values—that is,

$$LR = d_{dw}/d_{iw} = EC_{iw}/EC_{dw} \qquad \text{(B4.12.3)}$$

Thus, LR is defined as the fraction of irrigation water (of conductivity EC_{iw}) that must drain out of the root zone to maintain the EC in that zone (EC_{dw}) at or below a critical value. Experiments indicate the critical EC_e for *Vitus vinifera's* tolerance of salinity is 1.8 dS/m (see "Soil Testing and Salinity," chapter 3), which is approximately equal to an EC_{dw} of 3.6 dS/m. Hence, for irrigation water of 0.8 dS/m, we can calculate the LR as

$$LR = \frac{0.8}{3.6} = 0.22 \qquad \text{(B4.12.4)}$$

For vines on a salt-tolerant rootstock, the LR would be smaller because the critical EC_e is higher. Conversely, for the same LR, water of higher EC_{iw} could be used.

If one assumes no effective winter rainfall, this calculation shows that about one fifth of the irrigation water should drain below the root zone to avoid salt accumulation. In this case, for a vine water requirement of 300 mm during the growing season, the total amount of irrigation required to satisfy the vines and to provide an LR would be

$$d_{iw} = \frac{300}{1 - 0.22} = 385 \text{ mm} \qquad \text{(B4.12.5)}$$

However, the additional 85 mm of drainage would normally be provided by winter rainfall, so only when winter rainfall is low would additional irrigation be required during winter to provide the LR. The value of EC_{dw} should be monitored over several seasons to check that leaching is being effective in maintaining a salt balance.

Overall Salinity Management

Good salinity management must consider not only the soil's salt balance, but also the offsite effects of salt. Unnecessary leaching of salts into drainage water creates salinity problems for users downstream. "Best-practice" irrigation must therefore focus on the dual objective of minimizing saline return flows to streams, and avoiding salt buildup in the soil and in groundwater immediately below the soil.

Best-practice salinity management under irrigation involves the following:

- Keeping the *LR* as low as possible (≤0.1) by using low-*EC* water wherever available. This could involve blending good-quality water with high-*EC* water.
- Using salt-tolerant rootstocks when appropriate (see chapter 5)
- Matching the application of water to the vines' demand as assessed visually, from *ET* data, or by monitoring the soil water content
- Monitoring drainage so that *LF* is as close as possible to *LR* (see "Irrigation Methods")

Irrigation Methods

Irrigation methods for vineyards fall into two broad categories:

1. Microirrigation using drippers, microjets, or minisprinklers
2. Macroirrigation using overhead sprinklers, flood or furrow systems

Topography and soil type influence the choice of system, as well as the availability and price of water, the capital costs of installation, and the skill level of vineyard staff.

Microirrigation systems are all undercanopy methods of applying water. They offer the best means of accurately applying predetermined amounts at known rates, targeted to specific areas with minimum losses. Hence, the efficiency of water use from pump to target area can be as high as 80%–90% for well-managed systems. Although easily automated and controllable by wireless signals, these systems are the most capital intensive.

Macroirrigation systems are the least efficient, ranging from about 60% for fixed overhead sprinklers to less than 50% for furrow and flood systems. The latter tend to be found in older, established vineyards because of their low installation cost. However, they have become less viable with the increasing cost of water (resulting from scarcity) and because they are time-consuming to operate (and hence have high labor costs).

Microirrigation Systems

Drip Systems

Drip systems consist of emitters (drippers) spaced at regular intervals along flexible PVC pipes, usually suspended 0.3–0.5 m above the soil under the vines (although subsurface drippers are being more widely installed now in an effort to save water).

Emitter spacing and discharge rate depend on the vine spacing and soil texture, because the aim, particularly in dry regions, is to have a continuous band of moist soil in the rows. Specialist texts such as Nicholas (2004) and Pritchard and others (2004) describe details of the types of dripper available and installation methods.

The wetting pattern under an emitter changes with soil texture. The suction force pulling water sideways and downward is small in sand and sandy loams because of their large pores, and gravity predominates (figure 4.24A). In such soils the emitters can be as close as 0.5 m and, given the high infiltration rate, the discharge rate can be as much as 4 L/hour. As the texture changes to clay loam or clay,

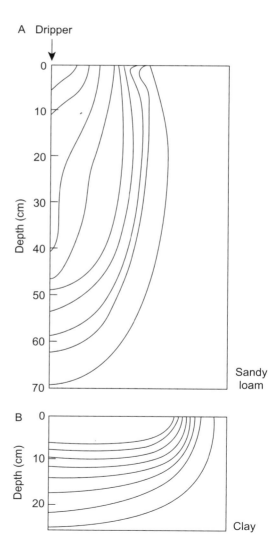

Figure 4.24 (A) Pattern of wetting under a dripper in sandy soil of high permeability. (B) Pattern of wetting under a dripper in clay soil of low permeability. (Redrawn from White, 2006)

the suction force pulling water sideways increases relative to the constant force of gravity so that the wetted zone is more like a flattened bulb (figure 4.24B). Also, because infiltration rates are slow, the preferred emitter discharge rate is 0.8–2 L/hour and the emitters can be placed farther apart.

Table 4.6 shows the amount of water needed to deliver a specified RDI, depending on soil texture and emitter spacing. In practice, if growers are advised to control the water content of the top 0.4–0.6 m of soil only, pulsed application of water (30 minutes on/30 minutes off) may be necessary to prevent deeper wetting in sandy profiles.

Depending on soil depth and permeability, subsurface drip lines are placed 0.2–0.3 m belowground, close to the vine row. This method is favored on clay soils where suction is dominant in drawing water into the soil and deep drainage is not excessive, and when reclaimed water is used (see "Reclaimed Water," chapter 2).

Table 4.7 summarizes the advantages and disadvantages of drip irrigation relative to other systems.

Microjets and Minisprinklers

Microjets, minisprinklers, and undercanopy impact-sprinklers deliver water at much higher rates than inline drippers and wet a larger proportion of the ground area. Thus, they do not save as much water as drip systems, but they do give better water distribution on sandy soils, thereby allowing a cover crop to be grown in dry regions. To avoid blockages, microjets and minisprinklers require filtered water or water with a low concentration of suspended solids. Nicholas (2004) and Prichard and others (2004) give further details.

Table 4.6 **Calculated Water Requirements for One Regulated Deficit Irrigation (RDI) Application on Two Soils of Different Texture**

Soil texture and depth	Vine spacing (m)	Irrigation conditions	RDI application[a] (mm)	Total water required (kL/ha[b])	Water required if only one third of soil area is wetted (kL/ha)
Loamy sand, 0.6 m	1 × 2.5	4-L/hr emitters at 0.5 m spacing (8000 per hectare)	37.5	375	125
Clay loam, 0.6 m	1.5 × 3	2 L/hr at 1 m spacing (2222 per hectare)	54.3	543	181

[a] Assume the RDI water required is equal to readily available water plus one half deficit available water, derived from table 4.4.

[b] Remember, 1 mm = 1 L/m².

Table 4.7 **Advantages and Disadvantages of Drip Irrigation**

Advantages	Disadvantages
Water applied to individual vines can be closely controlled; particularly suited to regulated deficit irrigation and partial root-zone drying. With frequent application, soil water content (and suction) is maintained in a narrow range.	Water should be filtered to avoid blocking of emitters by solids, chemical precipitates, or microbial growths. Although root invasion and soil blockage can be a problem for subsurface drip irrigation (SSDI), frequent flushing can prevent this. Vacuum relief valves should be installed at the head of the system and at high points along the laterals.
Little loss by soil evaporation and runoff (especially for SSDI), because area of wet soil is small and rates of application are low; no foliar salt uptake as with overhead sprinklers	Salt accumulates at the edge of the wetted zone, especially if there is insufficient winter rain to leach salts (more serious with water > 0.75 dS/m).
Low drainage losses (hence low loss of nutrients and potential contaminants), especially if drainage is monitored	Rodents, other mammals, and birds seeking water can damage the soft, flexible pipes (not a problem for SSDI).
Targeted fertilizer application by fertigation (see chapter 3)	Less effective microclimate control than with overhead sprinklers, which can be used to avoid frost damage to vines at critical times
Allows use of water of greater salinity than with overhead sprinklers, or flood/furrow systems (where irrigation is less frequent and the soil dries between applications)	More expensive than overhead sprinklers or flood/furrow systems
Wastewater can be used because chance of pathogens contaminating foliage or fruit is minimal (none for SSDI)	Wastewater may need to be chlorinated to prevent excessive bacterial growth and blockages.
Low operating pressures (100–200 kPa) mean low pumping costs; restricted weed growth, especially with SSDI, and no constraints on access because of wet soil	
Effective on marginal soils and in difficult topography (stony soils, steep slopes)	

Macroirrigation Systems

Overhead Sprinklers

Because overhead sprinklers wet all the ground area, water in excess of plant needs may be applied in parts of the vineyard, leading to drainage losses. Evaporative loss during water application can also be high, especially on windy days, and salts can cause scorch as water on leaf surfaces evaporates (figure 4.25).

In frost-prone areas, overhead sprinklers can be used to raise humidity in the vineyard and to provide a water film on leaf surfaces when a frost is likely. The high

Figure 4.25 Example of salt scorch on leaves in a vineyard on the lower Darling River, New South Wales, Australia. (Photo courtesy of Dr. Tapas Biswas, South Australian Research and Development Institute, Adelaide, South Australia)

heat capacity of water and the latent heat released as it freezes help to prevent leaf temperatures falling below 0°C.

■   Flood and Furrow Irrigation

For flood irrigation, the midrow area across to the mounded vine rows on either side comprises an irrigation bay. Water flowing from a head channel into the bay covers the entire mid row. Sufficient water must be applied for soil at the far end of the bay to wet to a desired depth, which usually means that vines at the head of the bay receive too much water. This uneven distribution is worse in long bays on permeable soils—a problem compounded further by the presence of slight rises and hollows where water ponds and infiltration is greater. Leveling the land with laser-guided graders when the vineyard is established minimizes this effect.

In furrow irrigation, water is diverted from a head ditch into shallow channels on either side of a vine row (figure 4.26A) or sometimes into a single median channel (figure 4.26B). Although furrow irrigation suffers from the same disadvantages as flood, both have the advantage of low capital costs and permit the use of water that would require filtration if used for drip irrigation.

Furrow and flood irrigation are wasteful of water compared with drippers and sprinklers, and are being phased out in most irrigated vineyards around the world.

(A)

(B)

Figure 4.26 (A) Furrow irrigation of Sauvignon Blanc vines on a gravelly soil in the Rueda region, Spain. (B) Single-furrow irrigation between rows of 90-year-old Zinfandel vines in the Lodi District, California, USA.

Monitoring Deep Drainage

Although capacitance or TDR probes can be used to monitor changes in soil water content at depth, because water moves in response to a gradient in suction, gravity, and osmotic forces combined, drainage can occur when changes in water content are small and undetectable by these probes.

Tensiometers placed at the bottom of the root zone are best used to detect deep drainage. As shown in table 4.2, the matric suction at the depth of the tensiometer bulb is equal to the tensiometer reading (measured in kilo-Pascals) less the placement depth in decimeters (10-cm intervals). A matric suction in the range 0–1.0 kPa indicates that the soil is very nearly saturated and drainage is occurring.

The Full Stop detector is another instrument capable of detecting the movement of a wetting front at depth (see www.fullstop.com.au). The Full Stop is installed at two depths—usually at 0.3 and 0.5 m—under drip-irrigated vines. A float rises when the matric suction at the funnel rim falls to 2 kPa or less, and about 20 mL of drainage water has collected in the reservoir at the base. Thus, the Full Stop not only shows when a wetting front has arrived, but it also allows a sample of water to be collected for analysis. As with the SoluSAMPLER (see figure 4.22), nitrate concentrations can be measured to check on any leaching loss that might lead to groundwater contamination. Also, measurement of EC_{dw} (see box 4.12) at successive times allows an estimate to be made of the amount of drainage.

Summary

Aggregation, aeration, drainage, water storage, and soil strength are important factors affecting vine growth that depend on soil structure. Structure is an expression of how particles of clay, silt, and sand combine to form aggregates, and of the nature of the pore spaces, or porosity, thereby created.

A soil's porosity is occupied by water and air. When fully wet, the soil is saturated, and as it drains, air gradually replaces water in the widest pores, followed by increasingly narrow pores. As this occurs, the suction forces acting on the remaining water gradually increase.

Normally, within 1–2 days a soil drains to its upper limit of available water, which is called the *FC*. The lower limit of available water—the *PWP*—is arbitrarily set at a suction of 1500 kPa. A soil's structure is considered to be very good when the air-filled porosity amounts to 15% of the soil volume and the *AWC* is 20% or more.

A predominance of exchangeable Ca^{2+} ions on the clay particles ensures soil aggregates are stable and able to resist the disruptive forces created by raindrop impact and cultivation. Well-humified organic matter also contributes to aggregate stability in topsoils, but less so in subsoils. Aggregation is potentially unstable in sodic soils in which exchangeable Na^+ ions occupy more than 6% of the soil's *CEC*. Gypsum is recommended for amelioration of sodic soils, but sodicity is particularly intractable when it occurs in the subsoil.

Compaction in sodic subsoils is measured as an increase in soil strength. Soil strength, measured with a penetrometer, should ideally be between 1 and 2 MPa when the soil is at its *FC*. Such a condition provides a firm base for machinery traffic, promotes good drainage, and allows easy root penetration.

Suction forces and gravity drive water movement into and through a soil. The suction arising from surface tension forces is called matric suction; the concentration of dissolved salts determines the osmotic suction. Gravity is the dominant force when the soil is wet, especially when there are large cracks in which water can flow downward. Organic matter in the surface of dry sandy soils can cause water repellency that is only slowly alleviated as the soil wets up.

Evaporation from the soil surface and transpiration through the vines (together called evapotranspiration, or *ET*) create an *SWD*. As the *SWD* increases, water goes from being easily extracted by vines (the *RAW* component) to more difficult to extract (the *DAW* component). For the same depth of soil, the sum of *RAW* and *DAW* relative to *AWC* is greatest for sandy clay loams and clay loams, and least for heavy clays.

In dryland vineyards, the supply of *RAW* plus *DAW* depends on the balance between rainfall and *ET*, soil texture, and the depth of rooting. In irrigated vineyards, depth of rooting is less important because irrigation is usually intended to control the water content of the top 40–60 cm of soil only. In both cases, moderate water stress from fruit set to veraison enhances fruit quality and the flavor and aromas of the wines produced. Moderate water stress is achieved more consistently in irrigated vineyards through RDI. Partial root-zone drying, in which the soil on alternate sides of a vine experiences drying, can achieve a similar effect. Regulated deficit irrigation and PRD are most effectively practiced with drip irrigation.

Midrow cover crops and undervine mulch moderate soil temperatures and affect soil water storage and biological activity. Drainage is determined not only by slope but also by soil texture and structure, especially in the subsoil. Poor drainage can lead to waterlogging, with consequential undesirable effects on root growth and soil microbial activity. Drainage can be improved through the installation of perforated pipes at depth, provided the water can be moved offsite. Some drainage is essential to prevent salt buildup in irrigated soils, especially when water of *EC* more than 0.75 dS/m is used. Pumping may be necessary to keep the groundwater table below 2 m depth.

5

The Living Soil

The more we learn about what's happening in soil, the more we learn about life.

■ Quote from an elderly Japanese farmer in David Suzuki's *Earth Time* (1998)

Soil and the Carbon Cycle

Soil is the living "skin" of Earth's terrestrial ecosystem. Like skin, it is bombarded by the sun's radiation, wind, and rain, and abraded by all manner of objects scraping its surface. Unlike skin, after an initial phase of weathering (see chapter 1), soil grows primarily from the surface down as plant and animal residues are continually added to it. These organic residues nourish a diverse population of organisms, feeding on the dead residues and on each other, in an interminable cycle of growth, death, and decay, commonly called the carbon (C) cycle (figure 5.1).

Green plants use energy from the sun to make carbohydrates, and subsequently proteins, lipids (fats), and other complex molecules, for their own growth and reproduction. As plant tissue matures, biochemical changes take place that lead to senescence; leaves yellow and eventually fall. In a perennial plant such as the grapevine, leaves are shed in winter, but roots grow, age, and die all the time. Prunings may also be returned to the soil. Collectively, the aboveground plant material returned to the soil is called litter. Root material that is "sloughed off" and C compounds that leak from living roots constitute rhizo-C deposition.

When digesting and decomposing C residues (the substrate) in litter and the soil proper, organisms obtain energy and essential elements for growth. In so doing, in a healthy soil, they consume oxygen (O_2) and release carbon dioxide (CO_2) to the air below- and aboveground, thus completing the C cycle. Directly analogous to the process described for nitrogen (N) in chapter 3, C is said to be immobilized in the bodies of the soil organisms, and mineralized when it is

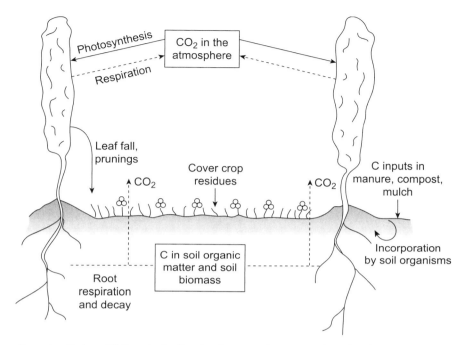

Figure 5.1 Carbon (C) flows in the C cycle of a vineyard. Inputs to the soil and vines, solid arrows; outputs as carbon dioxide (CO_2), dashed arrows.

released as CO_2. The proportion of C mineralized varies among the main groups of organisms, but the average for the whole soil is about 50%.

Individual organisms inevitably die and form the substrate for succeeding generations, which continue to decompose "old" and "new" substrates and release CO_2. The overall effect is a cascade, with CO_2 being released and organic matter steadily being transformed into more and more recalcitrant residues (figure 5.2). The combination of living organisms and dead organic matter, excluding living plant roots, is called soil organic matter (SOM).

Carbon turnover is the process whereby organic residues are continually added to soil and then decomposed. We can crudely measure the biological activity of a soil by its C turnover time, as described in box 5.1.

Disturbing the Carbon Cycle

When humans cultivate soil to grow crops, the C cycle is disturbed. Generally, the C released as CO_2, removed in harvested products, and lost by erosion and leaching, exceeds the input from photosynthesis. Because one consequence of organic matter decomposition is the mineralization of nutrients such as N, phosphorus (P),

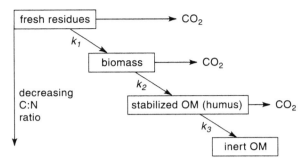

Figure 5.2 The "cascade effect" in the decomposition of organic matter (OM), starting from fresh plant residues. (White, 2003)

Box 5.1 Carbon Turnover in Soil

Steady-state equilibrium for soil carbon (C) is reached when the annual input of C in organic residues above- and below ground just balances the annual losses through decomposition, erosion, and leaching of dissolved organic matter. Assuming the loss of C by erosion and leaching is small relative to the loss by decomposition, we may write

$$\text{Loss (t/ha/year)} = -kC \qquad (B5.1.1)$$

Equation B5.1.1 shows that C loss is the product of a rate coefficient k and the amount C of soil carbon (by convention, loss is indicated by a negative sign). The coefficient k measures the fraction of soil C that decomposes during a given time period. For simplicity, if we chose a period of 1 year and a C input of A (measured in tonnes per hectare per year), we conclude that at steady state

$$A = kC \qquad (B5.1.2)$$

The coefficient k is a weighted average of all the k values for different classes of compounds, reflecting their ease of decomposition, as indicated in table B5.1.1. Thus, the average k value depends on the type of organic compounds present and their relative abundance.

As shown in figure 5.2, organic residues become more and more recalcitrant as they pass through successive stages of decomposition, so the k value of well-humified organic matter is much smaller than that of fresh plant material or most microbial residues. This explains why the net mineralization of nitrogen (N) from SOM is small relative to that from legume residues, as shown in table 3.2, even though the C-to-N ratio of SOM is quite favorable for N mineralization.

Rearranging equation B5.1.2, we see that

$$\frac{C}{A} = \frac{1}{k} \qquad (B5.1.3)$$

and because the units of k are per year, we have

(continued)

Box 5.1 *(continued)*

Table B5.1.1 **Organic Compounds and Their Relative Ease of Decomposition under Aerobic Conditions**

Compound	Ease of decomposition	End products
Sugars and starch, present in cell contents	Very easy; large k values	Carbon dioxide (CO_2), water (H_2O)
Proteins, present in cell contents and membranes	Very easy; large k values; less easily decomposed if complexed with polyphenols	Ammonium (NH_4^+) and sulfate (SO_4^{2-}) ions, CO_2, H_2O
Organic acids, present in cell contents	Easy to moderately easy; intermediate k values	CO_2, H_2O
Cellulose, cell wall material	Moderately easy to difficult, depending on the organisms present; intermediate k value	Humus, CO_2, H_2O
Fats and oils (lipids), cell contents and membranes; waxes on leaf surfaces	Difficult to very difficult; small to very small k values	Humus, CO_2, some orthophosphate ($H_2PO_4^-$) ions (from phospholipids), H_2O
Lignin and hemicellulose (cell walls); polyphenols (cell contents); chitin (insect exoskeletons, fungal cell walls)	Difficult to very difficult; small to very small k values	Humus, CO_2, H_2O

$$\frac{1}{k} = t \qquad\qquad\qquad (B5.1.4)$$

where t is defined as the C turnover time (in years). Thus, if the average k value for soil C decomposition is 0.05 per year, the turnover time is 20 years, which is theoretically the time taken for all the C in the soil to be replaced by incoming C. In reality, some of the fresh residues may last only a few weeks whereas other C that is the end product of many cycles of decomposition may last for 1000 years or more. Nevertheless, a soil with a C turnover time of 20 years or less is considered to be biologically active.

In a vineyard, the soil C equilibrium implicit in equation B5.1.2 will change if the variables A or k change. For example, rotary hoeing the mid rows will accelerate decomposition and lead to an overall decrease in SOM. Conversely, having a cover crop of permanent grass or a herbaceous mixture will decrease k and increase A, leading to a gradual increase in SOM. Cover crops can add from 1 to 10 t dry matter/ha, with the larger figure applying to fully organic/biodynamic vineyards. Compost used under the vines can add up to 2–3 t dry matter/ha.

and sulfur (S), a decrease in soil C resulting from cultivation and cropping gradually leads to a decline in soil fertility. There are also changes in soil structure.

Some traditional methods of viticulture have been developed to counteract this decline in C and soil fertility. For example, in the Champagne region of France, organic matter in the pale thin soils on chalk has for several centuries been enriched with lignite (called *cendres noires* or black ash) from local clay and silt deposits. During much of the 20th century, however, as chemical fertilizers were developed and used extensively in modern viticulture, emphasis on the role of SOM waned. Nevertheless, in more recent times there has been a swing back to organic methods in viticulture in most countries (see "Organic Viticulture," chapter 3). This chapter focuses on the benefits of maintaining SOM and a healthy diversity of soil organisms, and discusses the main biological problems of grape growing and their resolution.

The Soil Biomass

A variety of small organisms belonging to the plant kingdom—the decomposers—interact with many larger organisms, predominantly animals that live in and on the soil, called the reducers. Collectively, this host of organisms is called the soil biomass. The biomass is concentrated in the topsoil because that is where the food supply is most abundant, being made up of plant and animal residues falling on the soil surface, and C compounds released by roots.

The most important of the decomposers are bacteria, fungi (including the actinomycetes), and algae. There is also a group of very small animals—the protozoa, and the slightly larger nematodes—that do not fit neatly into either the decomposer or reducer categories because they feed predominantly on the true decomposers. Collectively, these small organisms, many of which cannot be seen with the naked eye, are referred to as the microbial biomass or, simply, as microorganisms. Measured in terms of C content, the microbial biomass amounts to 500–2000 kg/ha in the top 15 cm, with the larger number applying to soils under permanent grassland and forest.

In vineyards, the most important reducers are earthworms, wood lice, mites, springtails, millipedes, centipedes, and the adult and larval stages of insects. The relationships both within and between the reducer and decomposer groups are complex, but in the context of describing how organisms influence SOM turnover we may generalize as follows: First, although the reducers partially digest organic residues, their main benefit is in "reducing" or breaking litter and dead roots into small fragments that the decomposers then colonize and feed on. Earthworms, termites, and ants are examples of such reducers, but of these only earthworms are important in most vineyard regions. Second, life in the soil is very competitive and, during the course of evolution, organisms have evolved that feed on other organisms. Conceptually, we recognize a hierarchical arrangement of trophic or feeding levels with plant litter, dead roots, exudates, and detritus from organisms at the bottom. Successively, going up the "food web," there are

- Bacteria, fungi, actinomycetes, and algae that feed on dead organic matter
- Protozoa that feed on bacteria, fungi, and algae, and mites that feed on fungi
- Omnivorous nematodes and nematodes that feed specifically on bacteria, fungi, or other nematodes
- Top predators—mites and springtails—that feed on subordinate organisms

At any level in this hierarchy there are also organisms that feed on or parasitize living plant roots. Examples of such organisms that attack grapevines are discussed in subsequent sections. Symbiotic relationships also occur, such as the mutually beneficial fungal–root association described in "Mycorrhizas and Nutrient Uptake" in chapter 3, and the legume–*Rhizobium* symbiosis described later in this chapter.

Overall, a viable population of decomposers and reducers is essential for many soil processes such as energy flow (from dead to living matter and back again), nutrient cycling, disease control, and the maintenance of a stable soil structure. It is the cornerstone of a healthy soil. Box 5.2 outlines some simple criteria for recognizing a biologically healthy soil.

Box 5.2 Field Tests for a Biologically Healthy Soil

The following simple steps can be taken to assess a soil's biological health:

1. Scrape away any surface litter to reveal the top of the A horizon. A litter layer that is thin and well fragmented shows that organic residues are being broken down quickly and incorporated into the soil. The topsoil should be rich in humified organic matter (dark color), and the deeper this zone is, the better the soil's condition.
2. Observe the litter layer and soil surface for small insects such as wood lice and springtails. These should be plentiful and moving about actively.
3. Turn over a full spade depth of the topsoil and look for lively earthworms (see figure 5.4). More than 10 earthworm burrows per $10 \times 10 \times 10$-cm block of soil indicates good drainage and aeration (for roots and microorganisms).
4. Measure the initial weight of five dry strips of unbleached calico (5 cm long × 2 cm wide). Bury the strips in the topsoil, about 5 cm deep, and moisten the soil where the strips have been buried. Recover the strips after 3 weeks or so, wash them carefully and dry them. Reweigh the strips to measure their loss in weight and express this as a percentage of the initial weight. A loss of more than 10% in 3 weeks indicates a biologically active soil.

Functions of the Decomposers

Decomposer organisms vary greatly in their size, numbers, and diet. Table 5.1 provides a summary of the essential features of the main groups: bacteria, fungi, actinomycetes, and algae.

Among the decomposers, we recognize two broad nutritional groupings—the heterotrophs and autotrophs—as described in table 5.2. Irrespective of these differences, however, there are some generalizations we may make about modes of nutrition and metabolism within the decomposers.

First, microorganisms in soil are short of food most of the time, so when residues are added, the population quickly expands to consume the fresh substrate. Furthermore, if an unusual organic compound is added, organisms that can attack and consume that compound begin to proliferate because of their competitive advantage over other organisms that cannot.[a] When such a substrate is first added, there is a lag of several days or even weeks before it starts to disappear. However, once "enriched" with the requisite microorganisms, the soil quickly responds to further additions of that substrate, which is then rapidly decomposed. In vineyards, this behavior is most important for decomposing pesticides that can have an adverse effect on the environment because of their chemical stability or potential to leach into runoff water.

Second, soil microorganisms show a general response to drastic changes such as a rapid change from air dryness to wetness or from a frozen to a thawed state. When soil becomes air-dry or frozen, a substantial fraction of the microbial population dies and the remainder becomes quiescent. On rewetting or thawing, however, the bodies of the dead organisms provide an ideal substrate for the survivors. The population as a whole experiences a "flush" of activity that slowly subsides as the readily decomposable material is consumed.

Third, groups of microorganisms differ in their need for molecular O_2. Table 1.2 in chapter 1 introduced the distinction between aerobic and anaerobic conditions, but microbial metabolism is more complex in that we can recognize the following groups:

- Aerobic organisms that grow only in the presence of free O_2
- Facultative anaerobes, which normally use O_2, but can adapt to O_2-free conditions by using nitrate (NO_3^-) or other inorganic compounds as electron acceptors in respiration (see box 4.11), or C compounds as in fermentation
- Obligate anaerobes, which grow only in the absence of O_2 because free O_2 is toxic to them

---

[a] Because microorganisms such as bacteria have enormous "genetic plasticity," strains capable of consuming unusual substrates can evolve rapidly.

Table 5.1 **Essential Features of the Main Groups within the Decomposers**

Group of microorganisms	Growth habit	Reproduction	Numbers in the soil	Physiological characteristics
Bacteria	Single-cell organisms, 0.2–1 μm in size; favored by moist conditions	Rapid multiplication by cell division	1–4 billion/g soil; most species unidentified	Heterotrophic and autotrophic nutrition (see table 5.2); some parasites and symbionts; aerobes and anaerobes; collectively almost limitless variety in metabolism and ability to decompose diverse substrates
Fungi	Some (yeasts) are single celled, but most form filaments 1–10 μm in diameter (the hyphae) that collectively form a mycelium	Asexually by cell division and sexually by producing spores on fruiting bodies	<1 million/g soil; biomass is larger than bacteria	Heterotrophic; saprophytic and parasitic; some specialized symbionts as in lichens and mycorrhizas; more tolerant of soil pH < 5.5 than most bacteria and many can decompose recalcitrant C compounds (hemicellulose, lignin, and chitin)
Actinomycetes (Actinobacteria)	Grow as networks of fine hyphae less than 0.5 μm in diameter; more delicate than the fungi	Slow growing	No reliable numbers for soil	Heterotrophs; can decompose hemicellulose, lignin, and chitin; some parasites (pathogens); some "thermophilic" types grow in composts at temperatures more than 50°C
Algae	Filamentous or single celled; grow on land and in water	Asexual and sexual	100,000–1 million/g soil	Photosynthetic or heterotrophic (in the absence of light); Cyanobacteria or "blue-greens" fix atmospheric N_2 and prefer neutral to alkaline soils; green algae prefer acid soils

From Angle (2000) and Thorn (2000).

Table 5.2 **A Comparison of Heterotrophic and Autotrophic Microorganisms**

Heterotrophic microorganisms	Autotrophic microorganisms
Require carbon (C) in organic compounds to feed on and provide C for cell growth; include the majority of species of bacteria and all the fungi. Those that feed on dead organic matter are called saprophytes, whereas those feeding on living tissue are called parasites. Several parasitic bacteria and fungi are pathogens of grapevines (see "Fungal, Bacterial, and Viral Diseases" in this chapter).	Synthesize their cell substance from the C of carbon dioxide and include the remaining bacteria and most algae. Energy for this synthesis comes from sunlight (in the case of the photosynthetic bacteria and algae), or is chemical energy derived from the oxidation of inorganic compounds. The nitrifying bacteria are the most notable of the second type, one group of which gains energy from the oxidation of ammonium (NH_4^+) ions and a second from the oxidation of nitrite (NO_2^-), as described in box 3.2.

Figure 5.3 Mushroom fungi growing in a litter layer. (White, 2006; reprinted with permission of Wiley-Blackwell Publishing Ltd.)

Most fungi are aerobic, with a notable exception being yeast, which is a facultative anaerobe that ferments simple sugars predominantly to alcohol. Fungi and actinomycetes are more abundant in the litter layer because their ability to decompose lignin gives them a competitive advantage over bacteria (figure 5.3).

Lichens are a symbiotic association of a fungus and a photosynthetic alga (either a "blue-green" or "green" alga, or sometimes both). When a blue-green alga is one of the symbionts, the lichen can fix atmospheric N_2 and form protein (see table 5.1).

Fourth, soil microorganisms produce enzymes that act inside or outside their cells to digest substrates, but there are also enzymes that exist in soil independently of living organisms. Adsorption on soil mineral or organic matter shields these enzymes from denaturation and degradation. A common and stable soil enzyme is urease, which catalyzes the breakdown of urea in fertilizers and animal wastes (see box 3.2).

The versatility of soil microorganisms has encouraged attempts to culture specific groups for use as biofertilizers and biostimulants. Various organic products derived from the metabolism of SOM by microorganisms are also promoted as stimulants for plant growth. Box 5.3 discusses some of these products.

Box 5.3 Biofertilizers, Biostimulants, and Bioinoculants

In "Organic Viticulture" in chapter 3, we noted some of the organic products, commonly derived from seaweed, recommended for organic viticulture. These are usually promoted for their content of growth regulators (gibberellins, auxins, and cytokinins), sugars, and amino acids. There are also the special preparations used in biodynamic viticulture, obtained by incubating in soil a mixture of cow manure, quartz, and flower petals in cow's horn. Growers who use these products claim benefits for soil and vine health, grape quality, and the personality of the wines, especially their own enjoyment of them (see, for example, www.redwhiteandgreen.com.au). As discussed in chapter 6, this is part of the wider acceptance of a *terroir* effect and the production of single-vineyard wines. The products are best described as biostimulants for soil microorganisms, because the amounts added (as little as 90 g/ha) are far too small to have a direct effect on vine nutrition. Products derived from fishmeal are used as biofertilizers.

Humic acid in powder form (called humate) is recommended for improving soil *CEC* and soil structure, but to have any such effect it should be added in much larger amounts than most growers can afford. Although fulvic acid is promoted as a biostimulant that can penetrate leaves, its stimulatory effect is unproved. Humic and fulvic acids do not really exist as discrete organic compounds because they are a complex mixture of compounds, artificially derived from the chemical fractionation of SOM using strong alkalis and acids.

Many soil microorganisms can produce the plant hormone ethylene (C_2H_4). However, its actual production depends on soil aeration and substrate availability, with spore-forming bacteria living in anaerobic microsites being the primary producers. Ethylene concentrations as low as 10 parts per billion in the soil air can break seed dormancy and induce root and flower initiation, and leaf and fruit fall.

Various bacteria and fungi with specific functions have been isolated from soil. *Rhizobium* bacteria (for N_2 fixation in legumes) and mycorrhizal fungi (enhanced phosphorus (P) uptake by roots) have already been mentioned. Other microorganisms can mineralize organic phosphate or dissolve insoluble inorganic phosphates. An example of the former is *Bacillus megatherium*, present in the commercial inoculum phosphobacterin. Although some isolates have been effective in pure culture

(continued)

Box 5.3 *(continued)*

and pot trials in glasshouses, the beneficial effects are not consistent in the field. When introduced by seed inoculation into soil, such organisms are overwhelmed by competition from the much larger number of resident microbes. The introduced organisms do not survive or they lose their P-solubilizing activity when they reproduce.

Of the many attempts to manipulate the rhizosphere population by the introduction of specific organisms, few have produced consistent results except for the inoculation of legume seeds with *Rhizobium* (see box 5.4). However, recent research in New Zealand suggests that if mulch is inoculated with the fungus *Trichoderma*, the pathogen *Botrytis* can be suppressed. In Australia, a bioinoculant called Trich-A-Soil is available that contains selected strains of this fungus, but its effectiveness is unknown.

Figure 5.4 Example of shallow-burrowing earthworms collected from a moist, organic-rich soil.

Functions of the Reducers

Earthworms

Because of their size and physical activity, earthworms are more important than all the other soil invertebrates in achieving turnover of SOM in vineyards. As earthworms feed on dead organic matter they ingest large quantities of clay and silt-size particles and bacteria. Consequently, soil organic and mineral matter becomes more uniformly mixed and deposited in the worm feces, called casts. Sticky mucilage strengthens the casts, making them stable aggregates when dry that improve soil structure overall.

A few earthworm species live mainly in the surface soil and litter layer, provided that temperature and moisture conditions are favorable. Figure 5.4 shows

Table 5.3 **Earthworm Biomass under Different Management Practices in a Barossa Valley Vineyard**

Management practice	Earthworm biomass (kg/ha)
Mulch of grape marc	220
Ryegrass cover crop, slashed and thrown under the vines	300
Bare soil, surface application of lime	340
Straw mulch	450
Straw mulch plus lime	700

From White (2003); original data from Buckerfield and Webster (2001).

an example of one of these species. Others burrow more deeply and deposit their casts in horizontal burrows. Still others feed on surface litter and draw it deep into the soil, to be deposited in casts in their vertical burrows.

Studies in the Barossa Valley, South Australia, showed that the amount of organic matter available and the soil's chemical and physical condition influenced the species of earthworm present and their abundance. Consistent with studies on grasslands, it was found that most species prefer neutral or calcareous soils and few are found at a pH less than 4.5. Under hot, dry conditions, earthworms burrow as deeply as possible into the soil and aestivate. As shown in table 5.3, adding extra organic matter in the form of a straw mulch or a mown cover crop, and raising the soil pH, will increase earthworm numbers. In general, earthworms are most abundant under permanent grassland in cool, humid climates because there they are undisturbed, their food supply is plentiful, and their living conditions are most favorable. Numbers may then exceed 250 worms/m² or a staggering 2.5 million worms/ha.

Through earthworm activity, organic matter from the surface is mixed into the topsoil. Burrowing earthworms create pores, called biopores, as large as 5 mm in diameter that promote excellent drainage and aeration in the soil. Thus, soil with a healthy earthworm population quickly drains to *FC* after rain or irrigation.

Arthropods

The arthropods are a heterogeneous group of reducers that includes wood lice, mites and springtails, insects (beetles and their larvae, ants, bees, and wasps), centipedes, and millipedes. Many mites and springtails feed on plant residues and fungi in the litter, especially where thick mats build up under trees or undisturbed grassland. Predatory adult mites and springtails feed on other mites and nematodes, whereas the juveniles feed on bacteria and fungi. For example, the ability of predatory mites to control rust mite and bud mite is being studied in several wine regions in southeastern Australia. Such biological control methods can remove the need for chemical pesticides and hence reduce the environmental side effects of viticulture.

Wood lice prefer dark, damp places and are exclusively saprophytes. Many of the beetles and insect larvae that live in soil are saprophytes, but others feed on living tissues and so can be serious pests of grapevines. Bees, some wasps, and ants are important for the pollination of some plants used as cover crops. Centipedes are fast-moving, carnivorous arthropods, feeding on insects and other small animals, whereas millipedes are slow-moving plant-feeders, mostly saprophytic, although some feed on living roots, bulbs, and tubers and therefore can be pests.

Molluscs

Many molluscs (slugs and snails) feed on living plants and are therefore pests. Some species feed on fungi and the feces of other animals. Because the mollusc biomass is only 200–300 kg/ha in most soils, it makes only a limited contribution to the decomposition of organic matter.

Protozoa and Nematodes

Protozoa and nematodes (roundworms) are important because they control the size and composition of the bacterial and fungal populations, and in so doing influence the rate of nutrient turnover. Protozoa live exclusively in water films in very small pores. Next to the protozoa, the threadlike nematodes are the smallest of the soil fauna, ranging from 0.5 to 2 mm in length. Nematodes can number up to 10 million/m^2, mainly in the litter layer and topsoil, where they feed on fungi, bacteria, protozoa, and other nematodes. Root knot and root lesion nematodes are serious parasites of grapevine roots (see "Nematodes and Their Control" later in this chapter).

Building Up Soil Organic Matter

Microbial activity in soil is generally limited by the supply of substrate, so the addition of organic matter as plant residues, compost, and manure benefits the microbial population. Not surprisingly, microbial activity is four to five times greater in the immediate vicinity of plant roots (the rhizosphere) than in the bulk soil. Organic compounds secreted by the roots or cells sloughed off as the root grows through the soil are consumed by microorganisms, which in turn are consumed by other organisms. This rhizo-C fuels the prolific growth of microorganisms in the rhizosphere (discussed later).

The opportunity to build up SOM is greater in a well-managed vineyard than under most annual crops for which the soil is cultivated and much of the crop is harvested and removed. Many vineyard managers now focus on building up SOM through strategies ranging from conventional viticulture with minimum soil disturbance, to organic and biodynamic viticulture. Those who have adopted organic practices write with reverence about nourishing their soil, as the following example from an Australian Certified Organic vineyard in the Barossa Valley demonstrates:

> During autumn a cover crop of legumes (beans, vetch) and cereal (oats, triticale) is sown mid-row. This is mulched off and then later rotary-hoed into the ground as a green manure crop in spring. The green manure crop provides a rich source of organic matter and nutrition for the vines. The mixed crop also ensures a diverse array of soil fungi, bacteria and earthworms are present, providing a healthy biologically active soil. "Nutri-blend," a natural fertilizer consisting of humate, basalt and soft-rock phosphate inoculated with microbes, is also used in the vineyard. Beneficial insects such as ladybirds and spiders are abundant in the vineyard.... The overall philosophy is to get it right in the vineyard with a healthy and sustainable vine and soil balance to produce sound, flavoursome grapes. (Kalleske, 2007: 76)

The following sections explore the advantages and disadvantages of methods for building up SOM.

Cover Crops and Mulches

The effect of a midrow cover crop on the availability of soil water is discussed in "Cover Crops and Soil Water" in chapter 4. In addition to this effect, cover crops have important functions such as

- Protecting the soil from raindrop impact, which initiates structural breakdown and consequent erosion
- Improving soil structure, infiltration, soil strength, and drainage
- Building up SOM and enhancing biological activity
- Providing a habitat for beneficial insects and potential predators of grapevine pests
- Suppressing weeds

The relative importance of these functions depends on the soil type and the species composition of the cover crop. For example, a cover crop can be a single sown species, such as the barley crop shown in figure 4.18, or a mixture of species involving grasses and legumes, as seen in figure 5.5. Winter-growing annuals are preferred in regions that have substantial winter rainfall, but dry out severely in

Figure 5.5 A white clover–fescue cover crop late in the season in a vineyard on the Mornington Peninsula, Victoria, Australia.

summer. Alternatively, perennials may be preferred in regions where water availability is not a problem, but there is a need to control vine vigor, or prevent soil structural decline and erosion. In organic vineyards where biodiversity is highly valued, cover crops may consist not only of a mixture of grasses and legumes, but also of herbaceous broadleaf plants such as mustard (*Brassica* species), Compositae species, sages, buckwheat, and members of the carrot family.

Whatever the cover crop, mowing in early spring (at bud burst) is recommended to reduce the risk of frost damage to the vines. An annual cover crop, especially one containing a legume, can be cultivated into the soil in spring as a green manure (see "Organic Viticulture," chapter 3). A legume cover crop that is cultivated in when young and succulent will decompose rapidly, thus adding little to the SOM. However, its contribution to available soil N through mineralization can be appreciable, depending on how much N_2 the crop has fixed. Box 5.4 discusses the factors influencing N_2 fixation by legumes.

Cover crops compete with vines for water and nutrients. In some vineyards of the Bordeaux and Dordogne regions in France, for example, a cover crop is sown in every second mid row and alternated year by year (figure 5.6). Maintaining a cover crop during summer in low-rainfall regions is difficult where drip irrigation is used. Other possible disadvantages of a cover crop are that it may harbor pests such as the light-brown apple moth, although this is less of a problem with oats, faba beans, field peas, lupins, and mustard than with other broadleaf species.

Box 5.4 Factors Influencing Nitrogen Fixation by Legumes

Nitrogen (N) fixation begins when a legume root is invaded by *Rhizobium* bacteria (the endophyte) living in the soil. An infection thread grows down a root hair into the root cortex, where cells are stimulated to divide and enlarge to form a nodule (figure B5.4.1). Within a nodule, the bacteria change into bacteroids, which synthesize an enzyme called nitrogenase. The enzyme traps molecules of N_2 gas and reduces them to ammonia (NH_3), which is incorporated into amino acids.

The reduction requires a large amount of energy, generated from the metabolism of carbohydrates supplied to the endophyte by the host plant, which in return receives amino acids that combine to form proteins. Because it is beneficial to both the host and the endophyte, this association is called symbiotic.

Several factors determine the success of root hair infection, nodule initiation, and N_2 fixation. For example,

- Low pH and low calcium (Ca) supply inhibit root hair infection and nodule initiation
- Competition occurs between effective and ineffective *Rhizobium* strains at the time of root infection. An ineffective strain produces many nodules that do not fix N_2, yet their presence inhibits nodulation by an effective strain.

Figure B5.4.1 Nodules on the roots of a faba bean plant. Effective nodulation usually occurs on primary roots near the base of the stem.

(continued)

Box 5.4 *(continued)*

- Because of the large energy demand, N_2 fixation is less likely to occur when mineral N, especially ammonium (NH_4^+), is readily available
- An inadequate carbohydrate supply and water stress decrease a nodulated root's fixation capacity
- Nodulated legumes need more phosphorus (P), molybdenum (Mo), and copper (Cu) than non-nodulated legumes, and have a unique requirement for cobalt (Co)
- Nodulation and N_2 fixation can be stimulated by simultaneous infection of the roots by arbuscular mycorrhizas, mainly as a result of enhanced P supply to the plant

Actively fixing nodules are pink inside because of the pigment leghemoglobin that protects the enzyme sites from oxygen (O_2). *Rhizobium* strains show considerable variation in their effectiveness. When a legume requires a specific strain, it is cultured in peat and coated on the seeds with the aid of a sticky sugar solution. The seeds can also be pelleted with a coating of lime to improve the *Rhizobium's* survival in acid soils, and its subsequent infection of legume roots.

Because of these constraints and differences in host–bacterial strain performance, the quantity of N fixed is highly variable. However, provided an effective *Rhizobium* strain is present, a good guideline is that the amount of N fixed will be directly related to the proportion of legume in a cover crop. In vineyards where the cover crop occupies no more than two thirds of the land area, the amount of N fixed should be 50–100 kg N/ha/year. This fixed N is made available by nodules sloughing off and by legume residues decomposing in the soil, or through the excreta of stock animals if they graze in the vineyard. Note that this extra N may contribute to an excess vigor problem on more fertile soils.

There can also be a buildup of snails that attack young vines, and of parasitic nematodes.

Perennial cover crops are becoming more common in Australian vineyards, more so in the higher rainfall regions, with growth being controlled by mowing and/or herbicides, or by sheep grazing while the vines are dormant. Sometimes the cover crop may be allowed to spread under the vines, being controlled with herbicides or cultivation when necessary. Grasses with their fibrous root systems are better for improving soil structure than legumes and generally add more organic matter. Subsequent regeneration is ensured if the grasses are allowed to seed in early summer.

An undervine mulch created with mowings from a cover crop helps to suppress weeds and slowly builds up SOM. Heavy applications of mulch, such as

Figure 5.6 A mown grass cover crop in alternate rows in a vineyard near the Dordogne River, southwest France.

with cereal straw or compost, are the most effective in increasing SOM. A mulch of composted material should not need topping up for at least 3 years. Table 5.4 summarizes information on different types of cover crops and the plant species that can be used in vineyards.

Manures and Composts

When organic residues are consistently removed and not returned to soil, the soil becomes more acidic. Thus, adding manures and compost from elsewhere can counteract soil acidification in vineyards. The nutrient value of composted manures and other composts is discussed in "Cultural Practices for Organic Viticulture" in chapter 3.

Composting involves accelerating the decomposition of plant residues or farm animal manures in well-aerated, moist heaps. Nonindustrial sewage sludge (biosolids) can also be composted, provided bulking-up material such as wood chips or sawdust is added. Composting is done in heaps or windrows where the temperature should exceed 55°C for at least 3 days to kill weed seeds and pathogens. Depending on the nature of the raw materials and the ambient temperature, the process should be completed in 8–24 weeks. Well-made compost should be

Table 5.4 **The Range of Species Used in Cover Crops**

Type of cover crop	Common species	Attributes	Soil factors	Limitations
Winter cereals	Oats	Suitable for green manuring or "mow and throw"; go well with legumes	Wide range of soil pH; tolerate waterlogging	Slow, early growth
	Barley	More green yield than other cereals; competitive with weeds; green manuring	Salt tolerant; prefers soil pH ($CaCl_2$) > 5.5	
	Cereal rye	Does well in cool regions; drought tolerant; green manuring	Wide range of soil pH; benefits soil structure	Straw slow to decompose
	Triticale	Competes well with weeds; green manuring	Well adapted to acid soils; tolerates low fertility	Straw slow to decompose
Annual legumes	Faba bean	Winter growing; nitrogen (N) fixing; deep taproot; green manuring; early maturing	Suited to loams and clays	Should be inoculated in acid soils; unsuitable for mowing
	Field bean	Winter growing; N fixing; green manuring	Suited to loams and clays	Shallow root system; not suitable for mowing
	Vetch	Winter–spring growth; green manuring and mowing; combines well with cereals	Wide range of soil pH	Hard seeded; may germinate when no longer needed
	Lupin	Winter growing; deep taproot; green manuring	Suited to sandy soil and low pH; intolerant of free lime	Requires a specific *Rhizobium* strain; susceptible to waterlogging
	Medics	Winter growing; drought tolerant; hard seeded; green manuring	Prefer neutral to alkaline soils	Susceptible to waterlogging

Annual grasses	Annual and Italian ryegrasses	Winter–spring growth; very competitive; fibrous root system, green manuring, and mowing; combine well with legumes	Tolerant of a range of soil conditions, except low fertility for Italian ryegrass	Annual ryegrass produces large quantities of allergenic pollen
Brassicas	Fodder radish, rape, and mustard	Winter growing; fodder radish has large taproot that breaks up hardpans; green manuring	Glucosinolates released from the roots inhibit parasitic nematodes	Large N requirement; need to be sprayed out or incorporated before seed set
Perennials	Perennial ryegrass, fescue, and cocksfoot	Suitable for cool, high-rainfall areas; mainly winter–spring growth; fibrous root system; green manuring; combine well with legumes	Prefer loams and clays; cocksfoot the most tolerant of acid soils	May compete too strongly with vines
	Strawberry clover and white clover	Winter–spring and summer (provided irrigation is supplied); N fixing; green manuring and mowing; combine well with perennial grasses	Strawberry clover tolerates adverse soil conditions better than white clover	

From Nicholas (2004), and McGourty and Reganold (2005).

(A)

(B)

Figure 5.7 (A) A large compost heap under cover in a Willamette Valley vineyard, Oregon, USA. Air is forced through the heap to improve aeration. (B) Well-made compost in windrows in Victoria, Australia.

odorless and have a C-to-N ratio about 20. Figure 5.7 shows examples of large-scale composting operations.

Recently, a comprehensive review of the effects of compost applied as an undervine mulch was completed for the New South Wales Department of Environment and Conservation (Campbell and Sharma, 2008). Provided the mulch layer is at least 10 cm thick, notable benefits were identified such as

- A reduction in soil temperature up to 2°C (applicable to hot inland areas)
- A decrease of up to 10% in irrigation water required
- A reduction of 60%–100% in weed growth, thereby decreasing the need for herbicides

- A reduction of up to 30% in chemical fertilizer use (depending on the compost's nutrient analysis)
- Increased SOM (benefiting soil structure and biological activity)

In Australia, compost should be certified according to the Australian Standard for Composts, Soil Conditioners and Mulches (AS 4454–1999).

Biological Control in Vineyards

There are many soil-borne pests and disease organisms that affect grapevines directly or indirectly. Various chemicals are used to combat these pests and diseases, so it is important to know the fate of these chemicals and their residues in soil and the wider environment. Along with their attitude to many other products, consumers are concerned about (1) the impacts of grape growing and wine making on the soil, water quality, and human health; and (2) the "C footprint" of winegrowing (see chapter 6). Consequently, growers are interested in biological and less energy-intensive methods of controlling vineyard pests and disease; as well as the productive use of waste products, as discussed in "Winery Waste and the Soil" later in this chapter. Rootstocks offer one of the most effective methods of biological control in vineyards.

Rootstocks

Varieties of *Vitis vinifera* (the scion) can be grafted onto rootstocks that are better adapted to particular soil and site conditions. Initially, the prime purpose of such grafting was to enable vinifera varieties to resist the attack of phylloxera. This pest, endemic in North America, was accidentally introduced into France in 1862, nearly devastating the wine industry and subsequently having a serious effect on wine production in many other countries. Native American *Vitis* species such as *V. rupestris* and *V. riparia* that are resistant to phylloxera were investigated as rootstocks for grafting, mainly at Montpellier in France, where a statue commemorates the rescue of the ailing French vines by vigorous American rootstocks (figure 5.8). Other native American species such as *V. champini* and *V. berlandieri* were subsequently introduced into the grafting program, and many crosses have been made using these foundation species.

The four species commonly used for rootstocks have evolved in distinct regions of North America and hence vary in many of their characteristics other than their resistance to phylloxera. Not surprisingly, therefore, rootstock crosses have been developed that confer on the grafted vine differences in nematode resistance, salt and drought tolerance, susceptibility to lime-induced chlorosis, vigor on high-potential sites, nutrient uptake, yield, and fruit quality. These crosses are the basis of the commercial rootstocks now available. Although no

Figure 5.8 Memorial at Montpellier, France, to the rescue from phylloxera of French *Vitis vinifera* vines by resistant American rootstocks.

one rootstock will satisfy all requirements, the available range contains individuals most suited to particular conditions, as summarized in table 5.5 and discussed further here.

Growers should always consider using a rootstock

- When there is a risk of phylloxera attack
- When parasitic nematodes are present in the soil
- In salt-affected soil
- In drought-prone soil
- In a replant situation

Phylloxera and Its Control

Phylloxera (scientific name *Daktulosphairia vitifoliae*) is a small aphid that feeds on vine roots and is occasionally found in leaf galls. The first aboveground symptom of attack is a premature yellowing of leaves near harvest, which is more obvious when the vine is water stressed. Infected roots swell and become club-like,

Table 5.5 **Rootstock Crosses and Their Main Attributes**

Rootstock cross	Main commercial varieties	Phylloxera resistance	Nematode resistance	Other attributes
Vitis riparia × *V. rupestris*	3309 and 3306 Couderc, 101–14, Schwarzmann	Good	Variable, according to the species of nematode	Low to moderate vigor; not suited to dry conditions; Schwarzmann widely used in Australia in cool, moist regions
V. berlandieri × *V. riparia*	SO4, 5BB Kober/5A Teleki, 5C Teleki, 420A	Good	Moderate to high	Good affinity with *V. vinifera*, moderate to high vigor; suited to cool regions because of early maturity
V. berlandieri × *V. rupestris*	1103 Paulsen, 99 and 110 Richter (R), 140 Ruggeri (Ru)	High	Moderate to high, especially for 140 Ru	Vigorous; preferred in drier Mediterranean regions, 110 R and 140 Ru are very drought tolerant
V. champanii (natural hybrid of *V. rupestris* and *V. candicans*)	Ramsey, Dog Ridge	Good	Very high	Best adapted to hotter climates and deep, sandy soils; vigorous and drought tolerant; Ramsey is widely used in inland Australia
V. vinifera crosses with *V. berlandieri, V. rupestris*	Fercal, 1202 and 1613 Couderc, ARG 1, 41 B	Poor	Poor to moderate	Not recommended except in special situations in which lime tolerance is most important (use Fercal or 41 B)

Compiled from Whiting (2003).

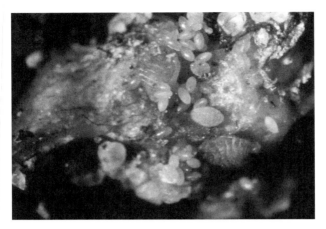

Figure 5.9 Phylloxera adults and eggs infesting a *Vitis vinifera* root. Note the swelling and yellow galls. (White, 2003. Copyright © The State of Victoria, Department of Primary Industries. Reproduced by permission of the Department of Primary Industries and therefore not an official copy. Photo by Greg Buchanan.)

with many yellow galls (figure 5.9). Rootlet growth stops and the uptake of water and nutrients is impaired. Progressively, there is a decline in vigor, stunted growth, and finally vine death after several years.

Because phylloxera starts from a point source and spreads outward, the organisms are most easily found on mildly affected vines at the edge of an infestation, usually in the top 30 cm of soil. Although remote sensing can detect weakened areas, these must be checked by soil inspection in the later part of summer (phylloxera are dormant in winter and shelter under bark on the roots). The severity and rate of spread of an infestation depend on the virulence of the phylloxera strain, there being 83 known genotypes with different degrees of lethality. In Victoria, for example, where phylloxera has existed since 1877, the biggest range of genotypes occurs in the older infected areas. Less virulent genotypes may be present for 30+ years without inducing symptoms, even though the organisms are numerous on the vine roots.

Phylloxera is a greater problem in poorly drained, heavy clay soils than in sandy soils. Phylloxera probably suffers from desiccation in sandy soils. For example, in the Nagambie Lakes subregion of central Victoria, a small plot of Shiraz vines survived phylloxera at the end of the 19th century because they were growing on a sandy ridge in an alluvial plain of clay loam soils (figure 5.10). Conversely, in Languedoc-Roussillon, France, phylloxera is controlled in some 1500 ha of vines on own roots by flooding the soil for 3 months each year.

Because insecticides are ineffective in controlling phylloxera, the best method of long-term control is to use cultivars that are grafted on resistant rootstocks. Control in planting material (cuttings and rootlings) is achieved by hot-water

These SHIRAZ Vines were Planted in 1860 and are the Oldest in the World.

Figure 5.10 Old Shiraz vines that survived phylloxera on a sandy ridge in the Nagambie subregion, Victoria, Australia. (White, 2003)

treatment, which is also used to control nematodes, crown gall, and phytoplasma disease (see box 2.4, chapter 2). Rootstocks with *V. vinifera* parentage should not be used when strong resistance to phylloxera is required. Table 5.5 summarizes the phylloxera resistance of the more widely used rootstocks.

Nematodes and Their Control

Several nematode species feed on vine roots, causing malformations and necrosis. Infested vines suffer reduced root function and loss of vigor, without any specific aboveground symptoms. Nematode damage is more severe in sandy soils and, overall, nematodes are a more serious problem than phylloxera in Australian vineyards. The more common pest nematodes are described in box 5.5.

As discussed in "Preplanting" in chapter 2, fumigants are effective nematicides, especially in replant situations, or where crops such as potatoes or tomatoes that harbor nematodes have been grown for several years. However, rootstocks provide the most effective long-term control of nematodes, and differences in the resistance of the common hybrids are indicated in table 5.5. Because *V. champini* has a high resistance to the root knot nematode, the rootstock Ramsey became popular in the Sunraysia and Riverland regions of Australia. *Muscadinia rotundifolia*, a subgenus of *Vitis*, also has strong resistance to root knot and dagger nematodes, which led to the breeding of the VR hybrids (vinifera × rotundifolia).

Box 5.5 Common Nematode Pests of Grapevines

The most widely distributed nematode pest is the root knot nematode (genus *Meloidogyne*), which is found in most sandy soils in Australia, especially in Sunraysia and the Riverland. At least four species—*M. incognito. M. javanica, M. arenaria,* and *M. hapla*—are serious pests of grapevines in different parts of the world. The nematode invades a root, causing cells to swell and form a gall about 3 mm in diameter, in which the females breed and lay eggs. A single gall with a succession of females may survive for several years. Some cover crops act as alternate hosts for these nematodes.

The most common species of root lesion nematode (genus *Pratylenchus*) found in vineyards are *P. zeae* and *P. vulnus*. Like *Meloidogyne*, this nematode also lives inside the roots. It burrows through the tissues without producing galls, but does cause root lesions that predispose to the invasion of fungal pathogens such as *Phytophthora* (a root rot fungus).

The dagger nematode (genus *Xiphinema*) is larger than other nematodes and feeds outside the roots, causing swelling of the root tips. The most common species in vineyards worldwide is *X. index,* but *X. americana* also occurs. The dagger nematode carries the grape fanleaf virus.

The citrus nematode (*Tylenchulus semipenetrans*) also feeds outside the roots. It is restricted to a few plant genera—most important, citrus and grapevines—and so is common where these two crops are grown in proximity.

However, their use in phylloxera-infested areas is questionable because of their vinifera parentage.

Apart from the use of resistant rootstocks, nematode damage can be mitigated by good nursery hygiene that ensures cuttings and rootlets are nematode free, and any practice that stimulates root growth and improves nutrient uptake, such as the use of manures, fertilizers, and compost; and also by improving soil structure and alleviating compaction.

Other Rootstock Attributes

▪ Salt, Drought, and pH

Tolerance of salt and drought is becoming increasingly important in the choice of rootstocks, especially as climate change affects warmer and drier vineyard regions. The threshold salinity in the root zone above which most *V. vinifera* varieties are affected is 1.8 dS/m, expressed as the saturation extract value EC_e (see "Soil Testing for Salinity," chapter 3). As discussed in box 4.12, an EC_e of 1.8 dS/m is approximately equal to an EC_{dw} of 3.6 dS/m, the latter being measurable on solutions obtained from a SoluSAMPLER (see figure 4.22). Monitoring root zone salinity is important because wine imported into the European Union must not contain more than 394 and 606 mg/L sodium (Na) and chlorine (Cl), respectively. As shown in table 5.6, selection of an appropriate rootstock can provide greater tolerance of salinity.

Soil depth, texture, and the availability of irrigation determine the choice of rootstocks for drought tolerance. Given that *V. riparia* (the river grape) naturally prefers a cool, moist environment, crosses based on this species are best suited to deep soils under cool, moist conditions. On the other hand, *V. rupestris* (the rock grape) naturally occurs in harsh habitats on rocky soils, so that crosses based on this species are best suited to warm, dry climates and shallow soils. Table 5.7 groups the common rootstocks according to these characteristics.

Table 5.6 Tolerance of Grapevine Varieties to Salinity

Tolerance level and salinity threshold[a]	*Vitis vinifera* and common rootstocks
Sensitive, <3.6 dS/m	Most *V. vinifera* varieties on own roots, 1202 and 3309 Couderc, K51–40
Moderately sensitive ⎱ 3.6–6.6 Moderately tolerant ⎰ dS/m	420A, 99 and 110 Richter, SO 4, 5BB Kober, 5C Teleki Schwarzmann, 101–14, 140 Ruggeri
Tolerant, >6.6 dS/m	1103 Paulsen, Ramsey

[a] Measured as the root zone electrical conductivity EC_{dw}.
Compiled from Zhang et al. (2002), Whiting (2003, 2004), and Roberts and Cass (2007).

Table 5.7 Choice of Rootstocks for Drought Tolerance in Different Soil Types

Soil profile characteristics[a]	Vineyard water status	Recommended rootstocks
Soil depth <20 cm; sand, loam, or clay, including any root-impeding subsoil	Dryland	110 Richter (R), 140 Ruggeri (Ru), 1103 Paulsen
	Irrigated	110 R, 140 Ru, 1103 Paulsen, Ramsey
Soil depth 20–75 cm; sands, loams, or clays, with no root-impeding subsoil	Dryland	99 R and 110 R, 140 Ru, 1103 Paulsen, Ramsey, 5BB Kober
	Irrigated	99 R and 110 R, Ramsey, 5BB Kober, 5C Teleki, Schwarzmann, SO4, 420A, 101–14 (in loams and clays)
Soil depth >75 cm; uniform or gradational profile of sand, loam, or clay	Dryland	99 R and 110 R, 1103 Paulsen, Ramsey (in sand), 5BB Kober
	Irrigated	SO4, 101–14, 5C Teleki, Schwarzmann, 3306 and 3309 Couderc, 420A

[a] See box 1.1, chapter 1, for soil profile descriptors.
Compiled from Whiting (2003), and Roberts and Cass (2007).

With respect to soil pH constraints, low pH can be corrected by liming (see box 3.6). Although vines can grow on soils with a pH as low as 4, they may suffer nutritional stress, and hence reduced growth, through the effect of active aluminum (Al) and possibly manganese (Mn). On calcareous soils (pH 8 and above), lime-induced chlorosis (see "Micronutrient Fertilizers," chapter 3) is not easily avoided unless *V. vinifera* varieties on own roots or lime-tolerant rootstocks are used. Although rootstocks derived from *V. rupestris* and *V. riparia* are lime sensitive, *V. berlandieri*, being a native of the limestone hills in southwest Texas and New Mexico, is the most lime tolerant of the American species and confers this tolerance on its crosses. Of these crosses (see table 5.5), the ranking from most lime tolerant to least is 140 Ruggeri > 5BB Kober > 420 Millardet > 110 Richter > 1103 Paulsen > 99 Richter > SO4 > 5C Teleki. Rootstocks with a vinifera parentage such as 41B and Fercal are the most lime tolerant, but should not be used where phylloxera is a threat.

■   Vine Balance and Wine Quality
The overall aims in selecting a rootstock–scion combination are as follows:

1. To avoid or minimize the influence of adverse soil factors (e.g., phylloxera, nematodes, salt, and pH)
2. To achieve a balance between root and shoot growth consistent with the yield and grape quality objectives for the vineyard

Depending on site potential, various rootstock–scion combinations range widely in performance from "high vigor" to "low vigor." Generally, the most vigorous vines are those with the most extensive root systems and greatest drought tolerance. Thus, in table 5.7 the more vigorous rootstocks are recommended for dryland viticulture on shallow- to medium-depth soils of lighter texture, whereas the less vigorous rootstocks are recommended for deeper soils and more favorable growing conditions.

Another factor affecting vine vigor and hence vine balance is the uptake of major nutrients such as N and potassium (K). Some soils, such as those shown in figure 1.2, have high N availability through the natural process of mineralization, but this can be managed through the use of a grass cover crop and irrigation control (e.g., keeping the top 15–20 cm of soil dry to restrict N mineralization and the uptake of N by the vines). Rootstocks vary in their uptake of K, and this can be an important selection criterion, especially for deep sandy soils in inland Australia that are naturally high in available K. For example, rootstocks derived from *V. champanii* (Ramsey, Freedom, and Dog Ridge) accumulate more K than 110 Richter, 1103 Paulsen, 5C Teleki, 5BB Kober, and SO4.

Fungal, Bacterial, and Viral Diseases

Some grapevine disease organisms reside in soil, whereas for others soil management influences the severity of the disease, or the actions taken for disease control affect the soil. Maintaining a biodiverse cover crop and/or mulching is helpful, if not in eliminating disease, at least in controlling the incidence of disease to a degree that spraying programs can be much reduced or even made unnecessary. For example, bunch rot or gray mold (*Botrytis cinerea*) attacks the flowers and fruit, affecting yield and wine quality. Although not soil borne, the fungus survives on mummified fruits so that vineyard hygiene is important. Research in the Marlborough region, New Zealand, has shown that an undervine mulch of shredded office paper reduced the incidence of gray mold to the extent that spraying was not needed. Mulching seems to work by improving the decomposition of vine debris, increasing competition between soil microbes and the fungus, and strengthening the skins so that the grapes are more resistant to fungal attack.

Examples of where disease control affects soil properties include the spraying of vines with wettable sulfur (S) to control powdery mildew (*Uncinula necator*). Regular use of S over many years can decrease the topsoil pH (see "Sulfur," chapter 3). Another canopy disease, downy mildew (*Plasmopara viticola*), overwinters on leaves in the soil. Spores that are splashed onto the foliage by rain (>10 mm), when temperatures are more than 10°C for at least 24 hours, will germinate and infect the leaves. Downy mildew has traditionally been controlled by copper (Cu) sprays (Bordeaux mixture and other Cu-based fungicides), which has led to potentially toxic Cu concentrations in some vineyard topsoils of the Bordeaux region. Table 5.8 gives other examples of fungal, bacterial, and viral diseases of grapevines.

Winery Waste and the Soil

Types of Waste

With climate change, many growers are experiencing a decrease in water supply and greater competition for irrigation water from other users. Furthermore, stricter controls are being placed on the discharge of untreated winery wastewater into surface waters. Consequently, the dual incentives for treatment of wastewater are

1. To recover a substantial fraction of winery wastewater treated to a level suitable for irrigation
2. To remove as much of the waste organic matter as possible, which can then be composted and used in the vineyard

Table 5.8 Some Fungal, Bacterial, and Viral Diseases of Grapevines

Organism	Mode of attack and effects	Prevention and control
Oak root fungus (genus *Armillaria*)	Spreads from oak tree roots, causes decay and death	Remove dead roots after deep cultivation and fumigate the soil
Black foot disease (genus *Cylindrocarpon*)	Invades roots and kills young vines rapidly, especially in poorly drained soil	Maintain good soil drainage; treat cuttings with approved fungicide
Eutypa dieback or "dead arm" disease (genus *Eutypa*)	Spores are released from dead wood and infect fresh pruning wounds to form cankers; stunted and chlorotic spring growth	Remove all dead wood; use approved fungicidal sprays; treat pruning wounds with boron solution
Bot canker (genus *Botryosphaeria*)	As for *Eutypa*, but forms "wedge-shape" cankers in trunk and cordons	Remove all dead wood and use approved fungicidal sprays
Esca or "black measles" (*Phaemoniella chlamydospora*)	Soil-borne fungus that invades roots and attacks the conducting tissue; "black measles" on fruit resulting from toxins	Remove diseased wood; maintain well-nourished vines
Flavescence dorée or "grapevine yellows"	Caused by phytoplasma organisms, similar to bacteria, which invade the conducting tissues and kill young vines	Spread by a leaf hopper, so need to control the insect vector
Pierce's disease (*Xylella fastidiosa*)	A bacterium hosted by some grasses and sedges growing along streams; can cause vine death	A serious problem in southern California; transmitted by a glassy winged sharpshooter that needs to be controlled
Leaf roll virus(es)	Viral infection decreases yield, delays ripening, and results in poor-quality wine	Transmitted by a mealybug, so need to control the bug and remove infected plants; use virus-free planting material
Grape fanleaf virus	Infects through dagger nematode attack; leaves are yellow and malformed (fanlike), poor fruit set, and many seedless berries	Virus is long-lived in root debris; remove infected vines and fumigate; use virus-free planting material

Compiled from White (2003) and Gubler et al. (2006).

Box 5.6 Amounts of Organic Waste from Wineries and Its Beneficial Use

Depending on the efficiency of the winery, a crush of 1000 t grapes produces between 1 and 3 ML wastewater. On average, the wastewater comprises 99% water and 1% solids (dissolved and particulate), most of which is organic matter. So for every 1000 t grapes producing 2 ML wastewater, there is about 20 t readily decomposable organic matter that can be recovered, composted, and applied to the soil. Separating the solids and water provides a source of water for irrigation.

For small wineries (<500 t crush), a favored option for water treatment is filtration through a wetland—for example, a *Phragmites* reed bed that removes much of the organic matter. The filtered water can then be used for irrigating nonvineyard land, such as pastures or woodlots, or can be blended with good-quality water at a ratio of about 1:10 and used to irrigate vines. Such filtration beds are expected to last up to 15 years before needing to be regenerated.

For intermediate-size wineries, the FILTER system developed by the Australian Commonwealth Scientific and Industrial Research Organization (CSIRO) is a possibility (www.clw.csiro.au/publications/projects/projects20.pdf). Tile drains (perforated PVC pipe) are installed at a 1-m depth and 10-m spacing in the water disposal area, which is divided into bays for flood irrigation. Much of the organic matter is retained in the soil as water percolates to the drains. However, dissolved salts leach into the drainage water, which must be pumped into evaporation ponds. The soil under the FILTER system retains more than 80% of the wastewater nitrogen (N) and phosphorus (P). The amount of water that can be applied in a season depends mainly on the soil texture. Clay soils tend to seal with the organic matter applied and may only accept up to 250 mm (2.5 ML/ha), whereas sandy soils may accept up to 1500 mm (15 ML/ha). In the latter case, for a 1000-t winery, about 0.2 ha of land would be required.

Approximately 10% of the crush goes into marc so that 100 t marc is produced per 1000 t grapes crushed. Although this material is not so readily decomposed, it can be composted (see figure 3.16) and applied to the soil. In Australia, most of the marc in the large grape-growing regions is extracted to remove residual alcohol, and then is dried and pelleted for use as stock feed. As such, it has almost half the nutritive value of feed-quality barley.

Marc (crushed skins, seeds, and stems) is another form of winery waste in addition to wastewater. Whether the waste is in liquid or solid form, there is the potential for substantial amounts of organic matter to be returned to the soil. Box 5.6 gives examples of the amounts of organic matter that are available in liquid or solid form, and their beneficial use.

Soil Factors to Consider in the Disposal of Winery Waste

The general principle to follow is to reduce, recycle, and reuse. Implementing this principle involves minimizing water use, segregating waste streams of different strength (e.g., keeping storm water separate from wastewater), treating the water and solids recovered, reducing the amount of cleaning agents used in washing equipment, and choosing cleaning chemicals with low Na concentrations.

The variability of the waste stream is a problem for wastewater management. The biochemical oxygen demand (BOD) is a measure of the concentration of biodegradable organic matter in the water, and can range from 1000 to 12,000 mg/L. Box 5.6 outlines treatment methods that are suitable for small wineries. For much larger wineries (>10,000 t crushed), anaerobic settling ponds are first used to remove some of the organic matter and to reduce offensive odors through decomposition. Some of the organic N and phosphorus (P) settles in the sludge, and nitrate N is lost by denitrification. These ponds are followed by aerobic ponds that are kept aerated to promote a balance between algal and bacterial growth, achieving a further reduction in BOD, N, and P. The effluent water from these ponds can be released to woodlots or pasture. The sludge from the ponds can be composted and used to build up SOM. Treating wastewater to a sufficiently high standard for irrigation in vineyards is expensive, costing as much as $4000/ML in Australia.

Where wastewater is intended for land disposal, BOD is less of a limiting factor than the total dissolved salts (TDS) and sodium adsorption ratio (SAR) of the water (see box 4.5, chapter 4). This constraint applies equally to the use of reclaimed sewage water in vineyards. Table 5.9 gives the range of dissolved salt and nutrient concentrations in winery wastewater. Likewise, the TDS of reclaimed sewage water, measured by the EC, is usually between 1.5 and 3 dS/m.

Because sodium hydroxide (NaOH) is commonly used as a cleaning agent in a winery, the SAR of the wastewater is raised and, over time, this can create soil structural problems, either at the disposal site or where the water is used for irrigation. To counter this problem, gypsum should be applied to the soil at regular intervals. As the gypsum dissolves, Ca^{2+} ions are released and so the SAR of the

Table 5.9 **Composition of Wastewater Typical of a Large Winery**

Measured component	Value[a]
Biochemical oxygen demand	500–7500 mg/L
pH	5–7.5
Electrical conductivity	2–4 dS/m[b]
Total nitrogen (mainly organic)	25–50 mg/L
Total phosphorus (mainly organic)	5–20 mg/L
Calcium	40–450 mg/L
Magnesium	15–60 mg/L
Sodium	40–150 mg/L
Sodium adsorption ratio	0.5–5 (mmol charge)$^{1/2}$L$^{-1/2}$

[a] The variation occurs both within and between seasons.
[b] Multiply by 640 to give total dissolved salts in milligrams per liter.

Figure 5.11 Healthy vines grown under irrigation with reclaimed sewage water in McLaren Vale, South Australia. (Photo courtesy of Dr. Belinda Rawnsley, South Australian Research and Development Institute, Adelaide, South Australia)

soil solution is decreased, ideally to less than 3 (corresponding to an exchangeable sodium percentage $<$ 6). Another approach is to use K-based chemicals for cleaning and to recover much of the K as potassium tartrate (this requires the separated solids be sent away for treatment).

Additionally, wastewater contains a small concentration of cationic polymers that are used in wine making to remove suspended colloids. These polymers (possibly alum, a complex aluminium sulfate) can adversely affect the biological health of the soil. However, if the wastewater pH is adjusted to greater than 7 with lime, the polymers are precipitated; also, any zinc (Zn) and Cu contaminants are precipitated as hydroxides. Measurements of microbial activity at long-term disposal sites (with pastures more than 20 years old) show that soil health is not impaired, provided the wastewater is properly treated initially. Similarly, soil microbial activity in a vineyard irrigated with reclaimed sewage water since 1999 in McLaren Vale, South Australia, is better than in soil irrigated with "mains" water. Figure 5.11 shows an example of healthy vines that are drip irrigated with reclaimed water.

Leaching Requirement for Salt Control

When wastewater is used for productive irrigation, the amount applied will be determined by the *ET* demand of the crop (see "Evaporation and Transpiration," chapter 4). However, repeated applications of wastewater will gradually increase salts in the soil and restrict the range of plants that can be grown. To avoid this problem, a leaching requirement (*LR*) should be incorporated into the irrigation schedule, as discussed in box 4.12, chapter 4.

Summary

The C cycle of growth, death, and decay is essential for replenishing a soil's nutrient store. A host of soil microorganisms colonizes organic residues, deriving energy and nutrients for growth. During the process, CO_2 is released and the organic residues are converted into more and more recalcitrant humic compounds. The combination of undecomposed residues, dead organisms, and humic compounds is called SOM.

Among the living soil organisms we recognize the reducers and the decomposers. The former are the larger organisms such as earthworms, wood lice, mites, springtails, insects and their larvae, millipedes, and centipedes that, although partially digesting organic matter, serve the important function of breaking it down into small pieces that the decomposers can colonize. The latter consist of bacteria, fungi, actinomycetes, algae, nematodes, and protozoa, which collectively have an extraordinary range of ability in decomposing organic compounds, as well as exhibiting complex predator–prey relationships. Reducers and decomposers together compose the soil biomass, amounting to as much as 2000 to 4000 kg "live weight" per hectare. The decomposers are referred to as the microbial biomass, which ranges in size from 500 to 2000 kg C/ha to a 15-cm depth.

The food supply or substrate is the main factor controlling the size of the soil biomass. Organic residues fall as litter and animal excreta on the soil surface, and dead cells and exudates are deposited around plant roots in a zone called the rhizosphere. Hence, these are the zones where the organisms flourish. Drastic changes such as air-drying or freezing a soil kill a large number of organisms, but the survivors multiply rapidly when favorable conditions are restored because they have less competition. Microorganisms can also adapt to consuming unusual substrates added to soil, such as pesticides, because in doing so they have a competitive advantage over nonadapted organisms.

A thriving and diverse biomass is the key to a properly functioning soil ecosystem, with benefits to plants possible from growth-regulating compounds produced by specific organisms. Organic and biodynamic viticultural practices place great emphasis on the beneficial effects of a healthy soil microflora and fauna. The mycorrhizal symbiosis between a fungus and grapevine roots improves P uptake, and the legume–*Rhizobium* symbiosis "fixes" atmospheric N_2 in plants such as clovers, beans, peas, and medics that are sown as cover crops.

Healthy soil microflora and fauna are achieved by maintaining an adequate input of organic materials through mulches, manures, and compost. A mulch moderates diurnal soil temperature fluctuations, reduces soil evaporation, and encourages earthworms. Cover crops also create a favorable habitat for earthworms, contribute to SOM, and improve soil structure.

Grapevines suffer from pests such as phylloxera and nematodes that live for part or all of their life cycle in the soil. The best long-term control for both groups of pest is to plant *V. vinifera* varieties grafted onto hybrids of the resistant American rootstocks (*V. riparia*, *V. rupestris*, *V. champanii,* and *V. berlandieri*). Grafted vines are available that give greater tolerance of dry conditions, salinity, and lime-induced chlorosis, and also provide better vine balance on both high- and low-potential sites.

Substantial volumes of liquid and solid waste are produced from a winery. Separation of the organic solids (about 1%) from water (99%) in the waste stream and treatment of the water enables it to be reused for irrigation. Solids, including marc, can be composted and applied in the vineyard. Treatment of the waste stream involves lowering its *BOD*. If the wastewater is applied to land, the *TDS* and *SAR* need to be monitored. Leaching may be required to prevent a buildup of salts in the soil, and gypsum must be applied to prevent the soil from becoming sodic.

6

Putting It All Together

> You have only to drink wines from specifically different
> vineyards to know that the notion of terroir is true.

■ Edited extract, "The Age of Terroir," in *Daily Wine News* (December 2007),
from Robert Geddes (2007), *A Good Nose & Great Legs: The Art of Wine from
the Vine to the Table*

Soil and *Terroir*

Chapter 1 refers to the importance of *terroir* in defining the distinctive character
of wines produced in many of the premium wine regions of France and other Old
World countries. This concept is rooted in the tradition of viticulture in those
countries. Now there are New World wine makers and viticulturalists who have
warmed to the concept, such as well-known wine maker Geoffrey Grosset of Clare
Valley, South Australia, who said "a wine has a certain taste not just because of the
[grape] variety and vineyard management, but because of its place" (quoted by
Max Allen, 2003: 45). The "sense of place" referred to by Grosset embodies the
influence of site as determined by the local climate (including slope, aspect, cool-
air drainage, and so forth) and the soil.

A number of anecdotal explanations are given as to why a particular soil con-
veys a certain taste and flavor to a wine. For example, according to James Wilson
(1998), variations in manganese (Mn) content in the granites and schists of northern Beaujolais in France are thought to change the character of the best Cru du
Beaujolais wines made from the Gamay grape. John Livingston (1998) stated that
the volcanic parent material of soils in the Stag's Leap area, Napa Valley, California,
conferred a distinctive character on the Cabernet Sauvignon wines, an idea shared
by other wine makers whose vines grow on basalt soils in the Willamette Valley,
Oregon (Pinot Noir wines), and ancient Red Cambrian soils in the Heathcote
region, Victoria, Australia (Shiraz wines). Unfortunately, the scientific basis for
these influences has not been established, primarily because the range of factors

involved in *terroir* and their possible interactions is such that the isolation of a single factor effect is extremely difficult. This has led some, such as Warren Moran (2006), to dismiss the influence of site factors and express the contrary view that "*terroir* is a social construction... really it's the people who have learned to understand and make the most of their environment."

Nevertheless, the empirical evidence for a connection between "a place" and the taste and flavor of the wine from that place is undeniable, as attested to by many wine makers and wine writers. The best hypothesis to explain the effect of place, from the point of view of the soil and its surroundings, is that vines should experience some degree of stress as the berries are filling and ripening. This stress, as elaborated in the following sections, is most likely to be induced by the rate at which the vine can withdraw water from the soil (Seguin, 1986), but it may also be a nutritional stress.

Notwithstanding these comments about soil and *terroir*, the main objective of this book is to clarify the reader's understanding of "what makes a soil tick"—What are the important properties for growing healthy vines? How does a soil respond to its vineyard environment (including the weather)? And how can human intervention modify soil behavior and the performance of grapevines growing in the soil? To some extent, creating ideal soil conditions by relieving a cause of stress, whether physical or nutritional, may be counterproductive to the expression of a site's natural *terroir*. Balancing these conflicting interests will depend on the viticulturalist-cum-wine maker's personal perception of the value of *terroir*.

Is There an Ideal Soil?

Even though it invokes a circular argument, any description of an "ideal soil" must acknowledge the influence of the local environment in which the vines are grown, the landscape features, grape variety, style of wine to be made, and the potential profitability of the enterprise. With this proviso, I identify the following key properties with respect to an ideal soil:

- Soil depth
- Soil structure and water
- Soil strength
- Soil chemistry and nutrient supply
- Soil organisms

Soil Depth

The natural tendency of the long-lived *Vitis vinifera* to root deeply and extensively gives it access to a potentially large store of water and nutrients, provided there are no subsoil constraints (see chapter 3). In sandy and gravelly soils that are naturally low in nutrients, such as in the Médoc region of France, Margaret River in Western Australia, and the Wairau River plain, Marlborough region, New Zealand, the deeper the soil the better. However, such depth can be a disadvantage where soils are naturally fertile and rain is plentiful, as in the King Valley and parts of the Yarra Valley, Victoria, Australia, because vine growth is not in balance. Although large yields can be obtained, this is generally at the expense of fruit quality in the sense of the intensity of flavors, especially for red wines.

Soil depth is of less consequence in irrigated vineyards where the normal practice is to control the water content of the top 40–60 cm of soil only. This practice restricts the number of roots below this depth, so the effective soil depth is limited. Similarly, rooting depth is restricted in shallow soils on hard rock unless that rock is fractured naturally, or by deep ripping.

Examples of vines growing on shallow soils on fractured schist occur in the Central Otago region of New Zealand and parts of Languedoc-Roussillon, France (see figure 1.3). There are many examples around the world where vines perform well on shallow soil over limestone—the mid slopes of the Côte de Nuit and Côte de Beaune in Burgundy, France (see figure 1.5); the Terra Rossa of the Coonawarra, South Australia (figure 6.1); on the "starfish limestone" of St. Emilion in Bordeaux;

Figure 6.1 Profile of a shallow Terra Rossa on limestone in the Coonawarra region, South Australia. Note the solution "tongue" on the left side of the profile. Vines are in the background.

and the calcareous shale and limestone of the Monterey formation in Paso Robles, California. In St. Emilion and the Coonawarra, the shallow limestone has in past ages been enriched with wind-blown loess (see table 1.1), and the rate of release of stored water from soil and underlying limestone seems to ideally regulate water uptake for the production of high-quality fruit of intense flavors.

Soil Structure and Water

Soil structure and water interact to determine the ideal conditions for aeration, water storage, and drainage. Fundamental to understanding this interaction is the relationship between the amount of water and the energy with which it is held, expressed as the soil water suction (see figure 4.12). This relationship determines the ease of water extraction by vines, the extent of gas exchange between the soil and the atmosphere, soil water storage, and speed of drainage.

A soil at field capacity (*FC*) should have 10%–15% air-filled porosity and an available water capacity (*AWC*) of 20%–25% (see figure 4.2). This range of air-filled porosity allows free diffusion of oxygen (O_2) into the soil to maintain normal respiration by roots and microorganisms, and prevents too much carbon dioxide (CO_2) and other undesirable gases from accumulating in the soil. Furthermore, the soil should drain rapidly after rain or irrigation, releasing water from its large pores (macropores) that then fill with air. Such a soil does not become waterlogged (provided there is no groundwater within 2 m of the surface).

An *AWC* of 20%–25%, being the amount of water held between *FC* and the permanent wilting point (*PWP*), translates into potential water storage of 200–250 mm/m depth of soil. The product of *AWC* and soil depth (in meters) determines the plant available water (*PAW*) for a given soil profile. As shown in table 4.4, chapter 4, sandy clay loams and clay loams have the most available water per centimeter of depth. Consequently, *PAW* is greatest in deep sandy clay loams and deep clay loams.

For irrigated vines, the depth of soil that has large values for readily available water (*RAW*) and deficit available water (*DAW*) can be more important than the *PAW* value, depending on the variety and whether regulated deficit irrigation (RDI) is applied. For example, a white variety such as Chardonnay does not benefit from water stress and does best on a deep soil with large *RAW*, *DAW*, and *PAW*. On the other hand, a vigorous red variety such as Shiraz (Syrah) is better grown on a relatively shallow soil of smaller *RAW* and *DAW*, so that RDI can be used more effectively to manipulate grape quality (see "Managing Water" later in this chapter).

In addition to these aspects, soil aggregation and aggregate stability are important. Figure 4.3 in chapter 4 shows an ideal topsoil structure of predominantly crumblike aggregates, best formed when organic matter is naturally high (>2% organic carbon [C] or about 3.6% organic matter) and well humified. Soil organic

C can be maintained and even increased with substantial inputs from compost, mulch, and manures, or from a permanent grass cover crop.

The role of organic matter is less significant in the subsoil where the type of clay mineral and the exchangeable cations are more important (see box 4.3). Iron (Fe) oxides (identified by their red color) and aluminum (Al) oxides contribute to stable subsoil structures, as do clay minerals that have calcium ions (Ca^{2+}) as the dominant exchangeable cation. Figure 4.4 in chapter 4 shows the ideal subsoil structure of a Terra Rossa in the Coonawarra region, South Australia, consisting of large, porous aggregates stabilized by Fe oxides, with well-defined fissures between the aggregates. The emphasis on exchangeable Ca^{2+} is necessary because too much sodium ion (Na^+) destabilizes a soil's structure. A soil's sodicity is defined by its exchangeable Na^+ content, expressed as a percentage of the cation exchange capacity (*CEC*), with a value more than 6% being undesirable.

Soil Strength

Soil strength influences how easily roots can push through a soil and determines what load a soil can bear without structural damage, especially when wet. Soil strength, bulk density, and aggregate consistence or cohesion are interdependent. For example, bulk density (inversely related to porosity) increases as a soil becomes more compacted and, at the same time, its bulk strength, as measured by a penetrometer, increases. Soil strength for easy root penetration, but good load bearing, is ideally between 1 and 2 MPa when the soil is at its *FC*. The roots of *V. vinifera* can penetrate soil toward the upper end of this range.

Although a soil may have these desirable "bulk" characteristics, it may not form a friable tilth as it dries. The ideal condition is when the larger aggregates under moderate pressure break down into many smaller, stable aggregates, as seen in the upper sample in figure 6.2. Soil that is compacted through natural causes because of its sodicity and unstable structure, for example, or because of too much wheeled traffic when wet, does not break down into a friable tilth but remains massive (the lower sample in figure 6.2).

Soil Chemistry and Nutrient Supply

Ideally, a soil should provide a balanced supply of the essential nutrients (see table 3.1). Deficiencies and toxicities can be identified through soil testing. However, critical soil test values for "available" nutrients such as phosphorus (P), potassium (K), and sulfur (S) are not specifically for grapevines, but have been adopted from other orchard and agricultural crops, mainly because plant analysis has been more favored for grapevines, especially in established vineyards. These problems are corrected with fertilizers (including organic inputs), soil amendments (such as lime and gypsum), and in some cases by changing the soil pH.

Figure 6.2 Example of a friable topsoil (upper) and a compacted topsoil (lower).

A soil's supply of nitrogen (N) and K deserves special mention. For N, there must be an adequate supply for healthy growth, flowering, and fruit set, as well as a concentration in the grape juice around 300 mg N/L for successful fermentation. Mineral N is produced naturally through the mineralization of organic N, and table 3.2 in chapter 3, gives a qualitative estimate of the net N supply from different organic sources. When this supply is inadequate, as in sandy soils in relatively hot climates, it can be supplemented with fertilizers. However, soils of high organic C content and C-to-N ratios well below 25 can produce more mineral N than a vine needs for balanced growth; the condition of "excess vigor" ensues. The extent to which excess vigor is a problem depends on the grape variety and the style of wine to be made, being especially undesirable for the production of high-quality red wines because too much shading of the grapes during ripening retards the full development of color and phenolic compounds. On the other hand, a white grape variety such as Sauvignon Blanc needs a good supply of N and water to develop the full aromatic potential of the fruit, for the making of wines meant to be consumed without ageing. Wines made from other white varieties such as Riesling and Semillon benefit from ageing and the use of fruit harvested from less vigorous vines.

An imbalance in the N and K supply to vines can lead to symptoms of "false K deficiency" in spring. Conversely, in soils containing illitic clay minerals, the slow release of K^+ ions from within the clay minerals can result in too much K uptake by the grapes, resulting in high juice pH and color instability in red wines. Thus, especially for red grape varieties, the preferred soils have a low N mineralization potential and a small proportion of illitic minerals in the clay fraction.

Although the chemistry of the micronutrients is complex, one generalization with respect to the effect of pH is that the availability of Fe, Mn, zinc (Zn), and copper (Cu) decreases with an increase in soil pH, that boron (B) is available up to about pH 8, whereas molybdenum (Mo) availability is least at low pH. Another consistent trend is that active Al increases as the soil becomes more acidic, which can adversely affect root growth and P uptake. However, vines have survived, if not thrived, on soils of pH as low as 4 for many years (see, for example, figure 1.3b). By similar reasoning to that advanced for moderate water stress inducing a distinctive wine character (discussed later), the nutritional stress of a very acid soil may be responsible for the distinctive "minerality" or "flinty taste" of the wines produced, as has been noted for white wines from the very acid soils of the Rangen region in southern Alsace, France.

An aspect of soil chemistry that is not directly nutritional concerns salts. A high concentration of soluble salts is undesirable. As discussed in "Salt, Drought, and pH" in chapter 5, for own-rooted vines the threshold value of the "saturation extract" electrical conductivity (EC_e) is 1.8 dS/m, which is approximately equivalent to 3.6 dS/m in the soil solution in the lower root zone. Even though there are rootstocks more tolerant of salts than own-rooted vines, if irrigation is used, the aim should be to keep the soil EC below the threshold, because the salt concentration in the root zone will increase over time. In soils in which vines can root deeply, groundwater should not occur within 2 m of the surface because dissolved salts can rise into the root zone by capillary action.

Soil Organisms

An ideal soil has a diversity of organisms, ranging from bacteria and fungi to arthropods and earthworms that interactively decompose organic residues. Most of these organisms are saprophytes (i.e., they feed on dead organic matter) and are therefore crucial to nutrient cycling. However, they also contribute to aggregate formation and stabilization through secreted gums and mucilage. Adequate organic matter, as indicated earlier in "Soil Structure," is a prerequisite for a healthy soil biota. Box 5.2 described some simple tests to determine whether a soil is biologically healthy.

A healthy soil biota implies a population of organisms with activities that are mutually beneficial, which can suppress undesirable organisms and which are at the least not harmful to plants. This concept is one of the underlying premises of organic (including biodynamic) viticulture—a fast-growing movement in viticulture globally (discussed later). However, there is no evidence that parasitic nematodes and phylloxera can be suppressed under organic viticulture, so a vineyard soil should be tested for these pests and grafted vines used as a preventative measure (see chapter 5).

Managing the Soil for Specific Objectives

Although wine grapes are grown on a variety of soils, no one soil is likely to have values of all the key properties in the ideal range. Furthermore, the ideal values for some of these properties will inevitably be different according to the style of wine to be produced. Thus, growers should manage the soil in a way that enables wine makers to achieve their specific objectives of wine style and market price point. Soil amelioration, fertilizers, water management, and biological tools can be used to modify soil behavior and the vines' response to it. In so doing, the grower must also ensure that grape growing and wine making do not adversely affect the health of the soil and the environment.

In setting objectives, the question arises: Is the wine to be an expression of place (*terroir*) or is it to conform to a brand image or "house style" that is consistent from year to year? In the former case, soil variability occurring over distances of tens of meters is a potential boon, with subtle differences in soil properties conferring an individual personality on the wine produced from different blocks. This approach is more likely to be followed by small growers who produce distinctive wines of one variety from areas as small as 1 ha or less. Figure 6.3 shows one such vigneron who is passionate about making a distinctive Pinot Noir wine from small blocks of vines. However, we should also recognize that wines produced from larger regions, such as the Côte de Beaune or Côte de Nuit in Burgundy are distinctive in reflecting the full complexity of the *terroir*—climate, soil, variety, and the wine maker's intervention—which wine writer David Schildknecht calls a "strong *terroir*" effect.

The second approach to wine making is more likely to be taken by large wine companies with extensive estates where a blended wine is made from fruit produced on several soil types (figure 6.4). In some cases, wineries adopt both approaches, depending on the characteristics of a particular season's vintage. For example, Tim Smith (2007) of Chateau Tanunda in the Barossa Valley of South Australia has said "we blend different Shiraz wines to achieve a 'house style,' but wines within the blend are occasionally worthy of keeping separate to show their individuality." Single-vineyard wines and wines deemed to reflect a particular *terroir* are increasingly being promoted as offering the consumer a distinctive wine experience.

Irrespective of whether grapes are grown to make a *terroir* wine or a blended varietal wine (or *vin de cépage*), there are several broad issues confronting the wine

Figure 6.3 A vigneron who specializes in making Pinot Noir wine with *terroir* on the Mornington Peninsula, Victoria, Australia.

Figure 6.4 An extensive vineyard on a gravelly river plain in the Marlborough region, New Zealand. (Photo courtesy of Richard Merry, CSIRO Land and Water, Adelaide, South Australia)

industry that bear directly or indirectly on soil management. These issues relate to the "sustainability" of a viticultural enterprise in the face of changing consumer preferences and market demands, and the need to ensure that appropriate inputs are available and natural resources conserved. The main issues to be considered here are

- The role of precision viticulture
- Managing water
- The role of organic and biodynamic viticulture
- Climate change and its consequences
- Integrated production systems (or IPSs)

Precision Viticulture

Precision viticulture is a concept based on the premise that the productivity of a vineyard, and even of blocks within a vineyard, is inherently variable. Natural soil variability is an important contributor to the overall variability in vine performance. Precision viticulture provides a range of tools for a grower to make better informed management decisions, thereby gaining better control over the production system.

Chapter 1 discussed the causes of soil variability. Chapter 2 described methods based on remote and/or proximal sensing to delineate the spatial structure of this variability. By making such measurements at a high spatial resolution and incorporating the data into a geographic information system (GIS), a grower can identify zones within a vineyard that can be managed separately, from planting through to growing and ultimately harvesting. When key soil and climatic variables are combined into a model, similar to the local site index described in chapter 2, which is then linked to a GIS, the resulting map output can be interpreted as defining digital terroirs for the site. Figure 6.5 shows the output of such a model that is under development. Tony Proffitt and colleagues (2006) give more examples of digital maps of variability in soil properties and vine vigor in their book *Precision Viticulture*.

Chapters 3 through 5 described how the expression of soil variability through nutrient supply, structure, aeration, drainage, soil strength, water storage, salinity, sodicity, and the soil's biological condition can be recognized. Actions to remedy any onsite constraints and the offsite effects of grape growing and wine making are recommended.

Given the emphasis on managing water (see the next section), an important application of precision viticulture is in irrigation design and layout. Through the choice of dripper placement and delivery rate, combined with flow control valves, water can be differentially applied to parts of a vineyard according to the available water and drainage properties of the soils. Using a map of vineyard soil variation, *RAW, DAW,* and *PAW* values can be estimated from soil textures and soil depths, as given in table 4.4.

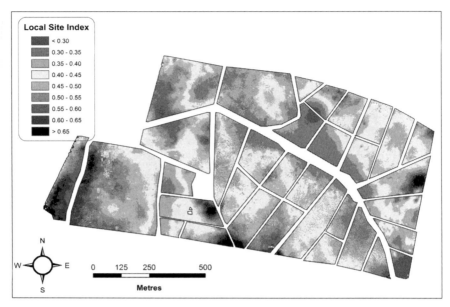

Figure 6.5 Map showing the digital terroirs for a vineyard in the Cowra region, New South Wales. The *terroirs* are defined by a local site index (LSI) incorporating site-specific net radiation, temperature differences resulting from aspect and height, soil readily available water (*RAW*), clay-to-silt ratio, and root zone depth. Sites with low LSI are better suited to late varieties because of lower temperatures and frost risk. These sites also have high *RAW* and are not suited to vigorous varieties like Shiraz. (Courtesy of Dr. James Taylor, Australian Centre for Precision Agriculture, Sydney, Australia)

None of these recommendations will convince a grower of the need to act unless they result in a better product, making more profit, with minimal environmental impact. To this end, the book *Precision Viticulture* gives examples where zones of different vine vigor within blocks were identified before harvest, and the fruit in these zones was separately harvested and vinified. Table 6.1 shows there can be substantial differences in the quality of wine produced, and hence its value, from such differential harvesting.

Some of the large wine estates and an increasing number of small producers are adopting this approach. For example, Lloyd Brothers' vineyard of 12.5 ha of Shiraz vines in McLaren Vale, South Australia, was planted in four blocks from 1998 to 2000 and originally treated as one commercial block with no regard to soil and other environmental variations across the blocks. Now, however, the top priority is to make small batch wines from selected rows within a single block. As Jodie Armstrong (2007) said, "The wines reflect the *terroir* of the slope with fruit from the top rows exhibiting a rich, ripe, dark berry character. The section in the

Table 6.1 **An Example of the Estimated Benefit of Selective Zone Harvesting in a Block of Premium Riesling Vines in the Clare Valley, South Australia**

Day of harvest	Fruit yield (t)	°Brix	Monetary value ($)[a]
1	15.9	22.0	28,548
2	15.2	21.6	27,450
3	10.3	22.1	18,594
8	12.0	21.4	21,564
9	17.2	22.0	30,906
Total	70.6		
Total income from separate harvesting			127,062
Additional harvesting costs			1568
Income from harvesting as a single parcel			70,600
Benefit from separate harvesting			54,894

[a] The fruit was valued at $1800/t when harvested separately compared with $1000/t when harvested as a single parcel.

Data from Bramley et al. (2005).

middle of the slope provides more juicy fruit flavours, with rows from the bottom of the block providing a perfumed blending agent. We typically view the vineyard in three sections: super-premium, premium and table wine."

Managing Water

Outside most of the AOC regions of France, vineyards have the option of managing the water supply to the vines to achieve yield and quality objectives. As described in chapter 4, such water management can be achieved through RDI or partial root zone drying (PRD).

Research in Australia, California, and Spain has shown that a soil water deficit (*SWD*) can have a positive effect on fruit quality, especially red grapes, but the size of the deficit (determining the degree of stress induced) and its timing must be appropriate. Expressed in terms of available water, the *SWD* should not exceed the sum of the soil's *RAW* and *DAW*. Alternatively, the soil matric suction should not exceed 100 kPa for a sandy soil or 400 kPa for a clay. In the Californian volume balance approach, the supply of water should not be less than 65% of full water use. Too large a deficit significantly reduces fruit yield, especially if prolonged to harvest.

With respect to timing, the three main periods of interest are from bud burst to fruit set, fruit set to veraison, and veraison to harvest. During the first period, flowering and fruit set are the most sensitive to water stress. An *SWD* greater than the soil's *RAW* causes significant yield reductions, primarily affecting the number of clusters per vine and the number of berries per cluster. Such stress does not normally occur in regions where the soil is at *FC* at the end of winter and is topped up by spring rains, when evaporation rates are still moderate. However, one effect of global climate change may be that *SWDs* exceeding a soil's *RAW* occur more frequently, especially in hot inland regions, so that irrigation will be needed early during the growing season to prevent a yield penalty. In this case, the *SWD* should be maintained within the soil's *RAW* range, to restrain early vegetative growth but avoid a subsequent serious yield reduction.

Regulated deficit irrigation that imposes moderate water stress on red varieties from the start of fruit set to veraison generally achieves the best outcome for fruit quality, although with some reduction in yield. Compared with fully watered vines, berry size decreases, soluble solids increase, malate concentration is lower, and total juice acidity is little changed. An increase in the concentration of phenolic (flavor) and color compounds is often associated with the decrease in berry size.

When RDI is prolonged through to harvest, although flavor and color compounds can increase, the yield reduction is often more pronounced and the concentration of soluble solids decreased. Recent research at the University of California at Davis has shown that varieties such as Cabernet Sauvignon, Merlot, and Zinfandel should be moderately stressed only until the berries reach 22–23° Brix, whereas for Syrah (Shiraz), the critical point is 18–20° Brix. Methoxypyrazine compounds that cause "green" flavors in wines reach a maximum before veraison and do not increase with subsequent watering. However, some growers with well-established vines on deep soils argue that "turning the taps off" after veraison is not a problem, provided that water stress does not cause shriveling of the mature berries and premature leaf fall.

Although PRD can effect similar improvements in fruit quality, the response is soil- and variety-dependent. Both RDI and PRD have the added advantage of saving water, which is a critical issue for irrigated viticulture in countries as widely separated as Spain, Australia, and South Africa. Savings are of the order of 30% for RDI and 50% for PRD. When drought occurs and water allocations for irrigation are much reduced, vines may receive only some 30%–40% of the optimum irrigation volume for one or more whole seasons. This is an example of sustained deficit irrigation, which may maintain fruit quality but will inevitably decrease yield and may have an adverse effect on the long-term sustainability of the vineyard.

Organic and Biodynamic Viticulture

Chapter 3 introduced the basic concepts and practices of organic viticulture. There are at least three main reasons for growers turning to organic viticulture:

1. Many wine writers and an increasing number of growers claim that organic wines taste better and "capture the *terroir* in the wine."

2. Organic viticulture benefits the soil because the inputs of organic materials and absence of chemical sprays promote a healthy soil biota, which in turn enhances vine growth and fruit quality.

3. Organic viticulture is good for the environment. Losses of nutrients by leaching are minimized, and C is conserved in the soil. Greenhouse gas emissions are indirectly reduced because chemicals and fertilizers that are made using fossil fuel energy are avoided.

Point 1 is a subjective opinion and difficult to separate from the marketing aspect of wine production. If consumers believe organic wines taste better, that is their choice. In particular, claims made for the benefits of biodynamic viticulture, which aligns its cultural practices with cosmic rhythms as advocated by Rudolph Steiner in 1924 (Steiner, 1993), are not testable by current scientific methods.

Point 2 has validity in the sense that inputs of organic materials generally do benefit the soil biota, which is important for improving soil structure and nutrient turnover (see chapter 5). The soil biota, especially earthworms, also benefit from not having potentially toxic accumulations of Cu in the soil from applications of Bordeaux spray over many years. Recent research supports the view that, with greater organic inputs and a reduction in spray use, the balance between predator and pest insects and disease in the vineyard is improved. However, whether these soil changes under organic viticulture measurably improve fruit quality and "capture the *terroir*" compared with conventional viticulture is debatable.

With respect to point 3, there are as yet no rigorous comparisons of the C balance between conventional and organic viticulture. Although some vineyards in France, New Zealand, Chile, and California claim to be "C neutral," this neutrality is achieved through C offsets, generally through paying for the planting of trees (buying C credits). However, more rigorous auditing needs to be done on both the vineyard C balance and the net effect of C sequestration in trees before claims of C neutrality can be justified. At the time of writing, version 1.0 of a Greenhouse Gas Accounting Protocol and Calculator has been released by the Winemakers' Federation of Australia for use by industry (see www.wfa.org.au). Produced by Provisor Pty Ltd of Australia under contract to wine industry organizations in California, Australia, South Africa, and New Zealand, this accounting tool is intended to become an international standard.

In Australia, an "organic or bio-dynamic farm must operate within a closed input system to the maximum extent possible. External farming inputs must be kept to a minimum and applied only on an 'as needs' basis," as permitted by the National Standard for Organic and Biodynamic Produce (2008: 50). Such production is often claimed to be sustainable. However, no commercial organic vineyard is sustainable in the long term if it is completely "closed," because nutrients are exported in the grape or wine products, and some loss occurs through leaching and as gases (in the case of N). A nutrient such as P gradually reverts to less and less soluble forms in the soil, gradually becoming unavailable to the vines.

Depending on the location, there is a small input of N from the atmosphere (see chapter 3) and a variable amount from legume N_2 fixation (see chapter 5), but overall nutrient export and loss will not be adequately matched by inputs of nutrients from the air, nor by the vines "mining" weathering rock in the subsoil. Thus, an external input of nutrients is needed for an organic vineyard to be biophysically sustainable, which means that some other part of the biosphere is being depleted of nutrients (and hence set to become unsustainable). Furthermore, if an organic vineyard is truly maintained as a closed system, the inevitable depletion of nutrients over a long period will result in unthrifty vines and poor yields, making the enterprise financially nonviable and therefore unsustainable.

Although recycling C and nutrients through the return of vine residues and composted materials is to be encouraged when possible, it needs to be done in conjunction with a calculation of the vineyard's nutrient budget, and soil and plant testing to assess the vines' nutrient status. Furthermore, it should be noted that the operating costs of organic viticulture are generally higher than for conventional viticulture, and organic growers in Australia at least are not paid a premium for their fruit by wineries.

Climate Change and Its Consequences

According to the report of the Intergovernmental Panel on Climate Change (IPCC, 2007), it is very likely that climate change resulting from human activities is occurring, with the main driver being increased emissions of the greenhouse gases CO_2, methane (CH_4), and nitrous oxide (N_2O). Because of the very large emissions from the burning of fossil fuels and land clearing, CO_2 is the most significant greenhouse gas, which is why most attention is being paid to reducing its emission (the "C footprint") and to increasing the size of C sinks in the biosphere.

Chapter 5 described the C cycle and reviewed methods of increasing soil organic matter (SOM). Increasing SOM in vineyards not only has a small positive effect on the global C sink, but it also has beneficial effects on such soil properties as aggregation and structure, the soil biota (its functions and diversity), nutrient cycling (N, P, and S in particular), and nutrient retention (through its effect on *CEC*). Through its effect on aggregation, organic matter helps to alleviate compaction in the topsoil. Figure 6.6 shows a vineyard in which SOM is augmented by means of prunings returned and a healthy midrow cover crop.

However, in warm to hot regions such as inland Australia, central and southern Spain, southern Italy, and the Central Valley of California, building up SOM is

Figure 6.6 A vineyard in central Victoria, Australia, showing prunings returned and a broadleaf midrow cover crop.

more difficult than in the cool-climate wine regions of northern France, the Rhine Valley, Piemonte in Italy, and coastal north and central California. Not only may supplies of straw mulch and compost be scarce and expensive in hotter and drier regions, but the residues also decompose faster in the soil than in cool regions. Although a grass cover crop can be used to increase SOM in hot, dry regions, maintaining the crop requires irrigation during the summer, which may impose an unacceptable extra demand for water. Figure 6.7 shows a well-grown grass cover crop in the Lodi district of the Central Valley, California, where water is supplied by subsurface drip irrigation to reduce evaporative losses. However, even in cool maritime climates such as Bordeaux, a combination of light textured soil and excessive cultivation can inhibit the buildup of SOM, as shown in figure 6.8.

The average global surface temperature increased by 0.6 ± 0.2°C during the past century, with the rate of increase being most rapid during the last two decades. The ±0.2°C indicates the degree of uncertainty about the average figure because of differences in measurement methods around the world. Similarly, uncertainty exists in predictions of future increases in temperature as a result of our incomplete knowledge of the rate of future greenhouse gas emissions and the sensitivity of the global climate to this perturbation. Notwithstanding the projections for overall global warming, of more relevance to the wine industry are the projections for regional changes in mean summer and winter temperatures, the number of frost-free days, the number of days with temperatures greater than 35°C, and changes in the amount and distribution of rainfall.

Analysis of growing season temperatures for some of the major wine regions in North America and Europe over the second half of the past century shows average warming trends ranging from 0.7°C in the Rhine Valley to 1.8°C in Bordeaux.

Figure 6.7 A well-grown grass cover crop in the Lodi district of the Central Valley, California, USA. Note the effect of subsurface drip irrigation on the left side of the mid row. (White, 2003)

Figure 6.8 Sandy, gravelly soil of low organic matter content in an overcultivated vineyard on the right bank of the Gironde River, Bordeaux region, France.

These differences in regional response to global warming are reflected in similar differences in the projected future responses, with the added complication of greater variability (the variance) in the average trend for any one region.

Thus, for southern Australia, the projected warming to the year 2030 is between 0.5 and 1.5°C, with concomitant decreases in the amount of winter and spring rainfall. Authors such as Leanne Webb and others (2006) and Greg Jones (2007) have discussed the likely impact of warming on the continued suitability of existing wine regions for their established varieties. The more narrow the temperature range most suited to the phenological development and ripening of a variety, the less likely it is to continue to flourish in its present region.

Apart from this effect, the projected decrease in winter and spring rain and increase in evaporation rates in southern Australia pose a serious challenge for winegrowing, one that underlines the importance of water conservation measures such as implementing efficient irrigation practices, reducing unnecessary evaporation, improving soil water-holding capacity, minimizing water waste in the winery, and using reclaimed water when possible. Indeed, water shortage is probably the greatest threat to the security of the Australian wine industry. In addition to competing for water with other rural water users and rapidly expanding cities, the industry is faced with a lack of opportunities for dam construction and the prospect of more frequent and intense droughts. Barrie Pittock (2005) gives an excellent review of the causes of global warming, its impacts, and mitigation of its effects in his book *Climate Change*.

Integrated Production Systems

In response to changing consumer perceptions and pressure from environmental regulators, wine industries in several countries have established IPSs for winegrowing. The overarching aim is to maintain the long-term sustainability of the wine industry.

Irrespective of country, the IPSs have very similar principles. One of the most comprehensive systems is the Sustainable Winegrowing Program (SWP), a cooperative venture between The Californian Wine Institute and the California Association of Winegrape Growers (see www.wineinstitute.org). The program defines sustainable winegrowing as "winegrowing and winemaking practices that are sensitive to the environment, responsive to the needs and interests of society-at-large, and are economically feasible to implement and maintain" (Wine Institute of California, 2005–2008). The interaction of these three principles of sustainability is often illustrated by a diagram such as that in figure 6.9.

All aspects of wine making are included, from soil and vineyard management, through pest control, energy efficiency, waste management and reduction, to wine quality, preferred purchasing, and working with neighbors. The Californian SWP is primarily educational, and participation is voluntary. By keeping detailed

Figure 6.9 The three overlapping sustainability circles. (Graphic courtesy of the California Sustainable Wine Growing Alliance)

records, growers and wine makers can assess their current practice against best management practices, enabling them to identify their strengths and weaknesses and so develop a plan for improvement.

Table 6.2 is an example of the performance categories for the organic matter criterion of soil management in the Californian SWP. Note the emphasis in table 6.2 on practices that lead to an increase in SOM and hence sustainability (categories 1–4), but this particular criterion can be "skipped" if organic matter is sufficient for the grower's soil type. The latter qualification illustrates the difficulty with all sustainability indices, such as SOM, that benchmark values have to be established for each environment and soil type, as was pointed out in "Climate Change and Its Consequences" earlier in this chapter. Establishing such benchmarks requires the collection and analysis of much data from a range of sites, but there is little evidence that such an activity is proceeding in a systematic way.

Examples of other countries operating IPSs are New Zealand (www. nzwine.com), South Africa (www.ipw.co.za), and Switzerland (www.vitiplus. ch/home-content.asp?siteid=vitiswiss). The underlying principles are similar to the Californian system in that self-assessment, education, and training are very important components. Although some systems have an auditing procedure and issue certificates of compliance, for others the existence and administration of environmental protection and food safety laws impose the necessary constraints on growers.

Table 6.2 **Performance Categories in the Organic Matter Criterion for Soil Management According to the Californian Sustainable Winegrowing Program**

Soil Management—Tilth				
Criterion	Category 1	Category 2	Category 3	Category 4
Organic matter	No organic matter is added to the soil other than what the vine produces, resident vegetation is minimized in the winter, and the vineyard is clean tilled	Resident vegetation is allowed to grow in the winter	Some form of organic matter is added to the soil annually (e.g., annual cover crop, compost, manure, or a combination of cover crop and manure or compost)	A combination of organic matter is added to the soil annually (e.g., permanent or annual cover crop, compost and/or manure), and tillage is reduced or eliminated to lower the rate of organic matter breakdown

Increasing sustainability ⇨

Compiled from Jordan et al. (2004).

Whichever IPS is operating, information about the soil and its management is essential to attaining the objective of sustainable wine production in the long term. Only when a grower takes care of the soil, together with the other components of production, can this aim be achieved. Furthermore, vineyard sustainability in the biophysical sense can only be achieved when there is a balance of inputs and outputs of essential nutrients, C, and salts; and avoidance of a buildup of toxic chemicals and pest organisms. This book provides the basic knowledge of how to attain biophysical sustainability, a necessary prerequisite to realizing sustainability in the fullest sense of the term.

Summary

Increasingly, wine writers and some wine makers are promoting the concept of *terroir*, and the soil's role in defining a wine's "sense of place." Although lacking scientific proof, a working hypothesis is that a constraint in soil water supply, and possibly nutrition, can confer a distinctive taste and flavor on wine produced at particular sites. However, this condition is not one that ensures optimum vine performance in terms of fruit yield and quality, and the profitability of the vineyard enterprise.

Five key factors for the optimization of vine performance are identified:

1. Soil depth
2. Soil structure and water
3. Soil strength
4. Soil chemistry and nutrient supply
5. Soil organisms

The optimum values for each of these properties depend on the vineyard environment, especially the local climate, the variety and style of wine to be made, and the wine maker's objectives.

In addition to the need to optimize vine performance, an overarching crucial issue for individual vineyards and the broader wine industry is the sustainability of viticulture. In the longer term, biophysical sustainability is a prerequisite of financial viability and the continued acceptance of grape growing by the community with which the industry coexists.

Progress toward biophysical sustainability can be enhanced by the application of precision viticulture, from site selection, vineyard layout, irrigation design, soil and vine management, through differential harvesting. The irrigation techniques of RDI and PRD offer opportunities for vigor control and improved fruit quality, as well as conservation of water, an increasingly scarce resource. The latter is a potential issue arising from global climate change, for which it is projected that some regions will become hotter and drier, but with more extreme weather events (e.g., a greater risk of late frosts and more very hot days during ripening). An additional example of the interrelatedness of these issues is that pest and disease incidence can change with climate change, as can the soil's ability to build up organic matter and sustain a healthy, diverse population of organisms. Avoiding the use of synthetic chemical pesticides, fungicides, and herbicides, and building up SOM to improve nutrient cycling and soil structure are two of the key tenets of organic and/or biodynamic viticulture.

In the quest for sustainability, several wine-producing countries have developed codes of best practice for viticulture and wine making as part of an IPS. The crux of such a system is that growing and wine making should be sensitive to the environment, responsive to the needs and interests of society, and economically feasible to implement and maintain—that is, to be environmentally, socially, and economically sustainable. The essential starting point in achieving this goal is good soil management in the vineyard that has minimal impact on the surrounding environment.

Appendices

Appendix 1 Conversion Factors for SI (System International) Units and Non-SI Units, Including American Units, and SI Abbreviations[a] Used in This Book

To convert column 1 into column 2, multiply by	Column 1, SI unit	Column 2, non-SI unit	To convert column 2 to column 1, multiply by
Length, area, and volume			
3.28	meter, m	foot, ft	0.304
39.4	meter, m	inch, in	0.0254
3.94×10^{-2}	millimeter, mm	inch, in	25.4
2.47	hectare, ha	acre, ac	0.405
0.265	liter, L	gallon	3.78
9.73×10^{-3}	cubic meter, m^3	acre-inch, acre-in	102.8
8.11×10^{-4}	cubic meter, m^3	acre-foot, acre-ft	1.233×10^3
35.3	cubic meter, m^3	cubic foot, ft^3	2.83×10^{-2}
0.811	megaliter, ML	acre-foot, acre-ft	1.233
Mass			
2.20×10^{-3}	gram, g	pound, lb	454
2.205	kilogram, kg	pound, lb	0.454
1.102	tonne, t	ton (U.S.), ton	0.907

(continued)

207

(continued)

To convert column 1 into column 2, multiply by	Column 1, SI unit	Column 2, non-SI unit	To convert column 2 to column 1, multiply by
Quantities per unit area			
0.893	kilogram/hectare, kg/ha	pound/acre, lb/ac	1.12
0.446	tonnes/hectare, t/ha	tons (U.S.)/ac, ton/ac	2.24
0.107	liter/hectare, L/ha	gallon/ac	9.35
Miscellaneous			
$(9/5°C) + 32$	Celsius, °C[b]	Fahrenheit, °F	$5/9 (°F - 32)$
9.90	megaPascal, MPa	atmosphere	0.101
10	Siemen/meter, S/m	millimho/centimeter, mmho/cm	0.1

[a] mega (M), $\times 10^6$; kilo (k), $\times 10^3$; deci (d), $\times 10^{-1}$; centi (c), $\times 10^{-2}$; milli (m), $\times 10^{-3}$; micro (μ), $\times 10^{-6}$; nano (n), $\times 10^{-9}$.

[b] To convert from degrees Celsius to degrees Kelvin, add 273.

Appendix 2 The Relationship between Electrical Conductivity 1 to 5 Soil-to-Water ($EC_{1:5}$) and Electrical Conductivity of the Saturation Extract (EC_e) for a Range of Soil Textures

To convert $EC_{1:5}$ values to EC_e values, multiply the $EC_{1:5}$ value by the factor K in the following table. The factor decreases as texture increases, because, per unit weight of soil, a clay holds more water at saturation than a sand. The dilution effect on the soil solution is therefore relatively smaller in a clay soil.

Note that these conversions are likely to overestimate EC_e in soil containing salts such as gypsum, because such salts dissolve to some extent when the soil is suspended 1 to 5 in water.

Table A2.1 **Factors to Convert $EC_{1:5}$ to EC_e Values According to Soil Texture**

Soil texture	Factor K
Loamy sand	13
Silty loam	12
Sandy loam, loam	11
Sandy clay loam, clay loam	9
Sandy clay, loamy clay, light clay	7
Medium to heavy clay	5

Adapted from White (2003).

Appendix 3 Calculating the Nitrogen Balance and Appropriate Amount of Nitrogen Fertilizer for Wine Grapes

The amount of fertilizer required is estimated from the nitrogen (N) balance equation

$$\text{N required} = \text{Inputs} - \text{Crop removal} - \text{Losses} \tag{A3.1}$$

Table A3.1 gives some indicative results using this approach, assuming a 10 t/ha grape yield.

The first example is typical of vineyards in southeastern Australia with no midrow cover crop, a soil carbon (C) content of 2%, and a C-to-N ratio of 12.5. Two rates of annual rainfall—600 mm and 1200 mm—are considered. The second example is for the same two vineyards, under the same two annual rainfalls, except that a legume cover crop is grown during the winter and cultivated in as a green manure in spring. The green manure adds 1.5 t C/ha to the soil at a C-to-N ratio of 20.

As shown in figure 3.14, N fertilizer can be applied at two growth stages, requiring about 2–3 g N per vine per week each time, depending on the vine density, as recommended for Western Australian vineyards. The range of 18–35 kg N/ha per season is similar to that used in the Bordeaux region, where an average of 30 kg N/ha per season is applied, and also for mature vines in the Willamette Valley of Oregon. However, the N input can be two to three times greater on young vines (up to 2 years old), and on sandy soils of low organic matter content. In all cases, the grower should be prepared to adjust the fertilizer input on the basis of plant analysis.

Table A3.1 **Examples of Nitrogen (N) Fertilizer Required (measured in kilograms nitrogen per hectare) Calculated from a Nitrogen Balance for a Crop of 10 t/ha**

Example 1: No cover crop

Atmospheric N input	Net soil mineralization	Crop requirement (shoots plus fruit) less recycling[a]	Estimated leaching plus gas losses[b]	Amount of N fertilizer required
5[c]	10	38	5	28
7.5[d]	10	38	15	35

Example 2: With a legume cover crop

Atmospheric input	Net soil mineralization, including green manure	Crop requirement (shoots plus fruit) less recycling[a]	Estimated leaching plus gas losses[b]	Amount of N fertilizer required
5[c]	25	38	10	18
7.5[d]	25	38	20	26

All figures on an annual basis.

[a] Average amount of N required for growth is 76 kg N/ha (range, 68–84 kg N/ha) (Mullins et al., 1992), less recycling of 50% through leaf fall and prunings.

[b] Depends on soil type and rainfall (plus any irrigation).

[c, d] Annual rainfall of 600 mm and 1200 mm, respectively. Atmospheric N inputs are apportioned 50% dissolved N and 50% dry deposition, except in the higher rainfall environment where dry deposition N decreases relative to dissolved N.

Appendix 4   Irrigation Scheduling Using Weather Data and Crop Coefficients for a Vineyard without a Cover Crop in Southeastern Australia

The aim is not to stress the vines from bud burst to fruit set (soil water deficit [SWD] kept within the readily available water [RAW] range), and to apply moderate stress from fruit set to harvest (SWD kept within the deficit available water [DAW] range). The effective rooting depth is 0.6 m. The soil has a loam over clay profile with RAW = 39 mm/0.6 m depth and DAW = 39 mm/0.6 m depth. Crop coefficients as given in table B4.8.1. All figures are rounded to the nearest whole number.

Stage of growth	Week no.	Precipitation, P (mm/week)	PET (mm/week)	Crop coefficient, C_c	Adjusted ET (mm/week)	P – ET (mm/week)	SWD (mm/0.6 m)	Drainage (–) or irrigation (+) (mm)
Initial state (FC)							0	
Bud burst (8 October)	1	10	15	0.2	3	7	0	–7
15 October	2	0	20	0.2	4	–4	–4	0
22 October	3	0	25	0.2	5	–5	–9	0
29 October	4	2	26	0.25	6	–4	–13	0
5 November	5	4	28	0.25	7	–3	–16	0
12 November	6	10	30	0.30	9	1	–15	0
19 November	7	0	36	0.35	13	–13	–28	0
26 November	8	0	38	0.35	13	–13	–41	0
Irrigate to return SWD to zero by applying 41 mm (41 L/m²)							0	+41
3 December	9	6	34	0.4	14	–8	–8	0
10 December	10	0	40	0.4	16	–16	–24	0
17 December	11	4	42	0.45	19	–15	–39	0
Irrigate to return SWD to zero by applying 39 mm (39 L/m²)							0	+39
Flowering (24 December)	12	0	46	0.45	21	–21	–21	0
31 December	13	0	49	0.6	29	–29	–50	0
Irrigate to return SWD to zero by applying 50 mm (50 L/m²)							0	+50
7 January	14	12	38	0.6	23	–11	–11	0
Fruit set (14 January)	15	5	40	0.6	24	–19	–30	0
21 January	16	0	49	0.6	29	–29	–59	0
28 January	17	0	46	0.3	14	–14	–73	0

(continued)

Stage of growth	Week no.	Precipitation, P (mm/week)	PET (mm/week)	Crop coefficient, C_c	Adjusted ET (mm/week)	P – ET (mm/week)	SWD (mm/0.6 m)	Drainage (–) or irrigation (+) (mm)
5 February	18	0	42	0.3	13	–13	–86	0
Irrigate to return SWD to –39 mm (upper DAW limit) by applying 47 mm (47 L/m²)							–39	+47
12 February	19	0	48	0.3	14	–14	–53	0
19 February	20	0	50	0.3	15	–15	–68	0
26 February	21	10	42	0.3	13	–3	–71	0
6 March	22	0	40	0.3	12	–12	–83	0
Irrigate to return SWD to –39 mm by applying 44 mm (44 L/m²)							–39	+44
Veraison (13 March)	23	0	38	0.3	11	–11	–50	0
20 March	24	0	35	0.3	10	–10	–60	0
27 March	25	5	30	0.3	9	–4	–64	0
3 April	26	0	28	0.3	8	–8	–72	0
10 April	27	6	32	0.3	10	–4	–76	0
Harvest (17 April)	28	0	26	0.3	8	–8	–84	0
Irrigate to return SWD to –39 mm by applying 45 mm (45 L/m²)							–39	+45
Postharvest (24 April)	29	12	24	0.3	7	+5	–34	0
Total irrigation water applied								266 mm or 266 L/m²

PET, potential evapotranspiration; ET, evapotranspiration; FC, field capacity.

References

Allen, M. 2003. Terroir Australis. *The Weekend Australian Magazine*, 13–14 December 2003, 45. Sydney: News Corporation Ltd.

Amerine, M. A., and A. J. Winkler. 1944. Composition and quality of musts and wines of California grapes. *Hilgardia* 15, 493–575.

Angle, J. S. 2000. Bacteria. In *Handbook of soil science*, ed. M. E. Sumner, C-14–C-22. Boca Raton, Fla.: CRC Press.

Armstrong, J. 2007. Winebiz feature of the week. *Daily Wine News* (www.winebiz.com. au). Adelaide: Winetitles.

Bramley, R. G. V., A. P. B. Proffitt, C. J. Hinze, B. Pearse, and R. P. Hamilton. 2005. Generating benefits from precision viticulture through selective harvesting. In *Proceedings of the 5th European Conference on Precision Agriculture*, ed. J. V. Stafford, 891–898. The Netherlands: Wageningen Academic Publishers.

Buckerfield, J. C., and K. A. Webster. 2001. Managing earthworms in vineyards: Improve incorporation of lime and gypsum. *The Australian Grapegrower and Winemaker* technical issue no. 449a, 55–61.

Campbell, A., and G. Sharma. 2008. Composted mulch for sustainable and productive viticulture. *The Australian and New Zealand Grapegrower and Winemaker* 530, 22–25.

Cass, A. 1999. Interpretation of some physical indicators for assessing soil physical fertility. In *Soil analysis: An interpretation manual*, ed. K. I. Peverill, L. A. Sparrow, and D. J. Reuter, 95–102. Melbourne: CSIRO Publications.

Cass, A., B. Cockcroft, and J. M. Tisdall. 1993. New approaches to vineyard and orchard soil preparation and management. In *Vineyard development and redevelopment*, ed. P. F. Hayes, 18–24. Adelaide: Australian Society of Viticulture and Oenology.

Cass, A., D. Maschmedt, and J. Chapman. 1998. Managing physical impediments to root growth. *The Australian Grapegrower and Winemaker*, June, 13–17.

Charlesworth, P. 2000. *Irrigation insights number 1: Soil water monitoring*. Canberra: Land and Water Australia.

Christensen, L. P. 2005. Foliar fertilization in vine mineral nutrient management programs. In *Soil environment and vine mineral nutrition*, ed. L. P. Christensen and D. R. Smart, 83–90. Davis, Calif.: American Society of Enology and Viticulture.

Christensen, L. P., A. N. Kasimatis, and F. L. Jensen. 1978. *Grapevine nutrition and fertilization in the San Joaquin Valley.* Oakland, Calif.: Division of Agriculture and Natural Resources, University of California.

Coates, C. 2005. *The great wines of France: France's top domaines and their wines*. London: Mitchell Beazley.

Emerson, W. W. 1991. Structural decline in soils, assessment and prevention. *Australian Journal of Soil Research* 29, 905–921.

Fertilizer Federation Industry of Australia. (ed.) 2006. *Australian soil fertility manual*. 3rd ed. Melbourne: CSIRO Publishing.

Geddes, R. 2007. *A good nose & great legs: The art of wine from the vine to the table*. Sydney: Murdoch Books.

Goldspink, B. H., and K. M. Howes. 2001. *Fertilisers for wine grapes*. 3rd ed., bulletin 4421. Perth: Agriculture Western Australia.

Goodwin, I. 1995. *Irrigation of vineyards*. Tatura: Agriculture Victoria.

Gubler, W. D., P. E. Rolshausen, F. P. Trouillas, G. M. Leavitt, and E. A. Weber. 2006. Grapevine diseases in California. In *Wine growing for the future: Proceedings of the 6th International Symposium for Cool Climate Viticulture and Oenology* [CD], ed. G. L. Creasy and G. F. Steans, KnWG8. Auckland: New Zealand Society for Viticulture and Oenology.

IPCC. 2007. Summary for policy makers. In *Climate change 2007: Synthesis report*. Fourth assessment report of the intergovernmental panel on climate change (www.ipcc.ch/ipccreports/ar4-syr.htm), 1–22. Geneva: WMO and UNEP.

Isbell, R. F. 1996. *The Australian soil classification*. Australian soil and land survey handbook. Melbourne: CSIRO Publishing.

Isbell, R. F. 2002. *The Australian soil classification*. Rev. ed. Australian soil and land survey handbooks series, vol. 4. Melbourne: CSIRO Publishing.

Jefford, A. 2002. *The new France: A complete guide to contemporary French wine*. London: Mitchell Beazley.

Jones, G. 2007. Climate change: Observation, projections, and general implications for viticulture and wine production. *Practical Winery and Vineyard* 29 (2), 45–63.

Jordan, A., J. Dlott, and K. Birdseye. 2004. The Californian wine community's code of sustainable winegrowing practices: From ground to bottle. In *Third Global Conference: Environmental Justice and Global Citizenship* (www.sureharvest. com), 1–9. Soquel, Calif.: SureHarvest.

Kalleske, T. 2007. Kalleske wines 2005 'Greenock' Barossa Valley Shiraz. *Australian and New Zealand Wine Industry Journal* 22 (3), 76–77.

Livingston, J. 1998. The geology of fine wine. *California Wild*. California Academy of Sciences Quarterly.

Maschmedt, D. J., R. W. Fitzpatrick, and A. Cass. 2002. *Key for identifying categories of vineyard soils in Australia.* CSIRO Land and Water, technical report no. 30/02. Adelaide: CSIRO Land and Water.

McDonald, R. D., R. F. Isbell, J. G. Speight, J. Walker, and M. S. Hopkins. 1990. *Australian soil and land survey field handbook.* 2nd ed. Melbourne: Inkata Press.

McGourty, G. T., and J. P. Reganold. 2005. Managing vineyard organic matter with cover crops. In *Soil environment and mineral nutrition,* ed. L. P. Christensen and D. R. Smart, 145–151. Davis, Calif.: American Society of Enology and Viticulture.

McKenzie, D. C. 2000. Soil survey options prior to vineyard design. *Australian Grapegrower and Winemaker* 438a, 144–151.

Moran, W. 2006. Quoted in R. Hamilton. Sun, slope and fog are part of a vineyard's terroir, but what about the owner? *Daily Wine News* (www.winebiz.com.au). Adelaide: Winetitles.

Mullins, M. G., A. Bouquet, and L. E. Williams 1992. *Biology of the grapevine.* Cambridge: Cambridge University Press.

National Standard for Organic and Biodynamic Produce. 2008. ed. 3.3 (www.daff. gov.au/agriculture-food/food/organic-biodynamic). Canberra: Australian Quarantine and Inspection Service.

Nicholas, P. (ed.) 2004. *Soil, irrigation and nutrition.* Grape production series no. 2. Adelaide: South Australian Research and Development Institute.

Pittock, A. B. 2005. *Climate change: Turning up the heat,* Collingwood, Victoria: CSIRO Publishing.

Prichard, T., B. Hanson, L. Schwank, P. Verdegaal, and R. Smith. 2004. *Deficit irrigation of quality winegrapes using micro-irrigation techniques.* Davis, Calif.: University of California Cooperative Extension, Department of Land, Air and Water Resources, University of California.

Prichard, T., and P. S. Verdegaal. 2001. *Effect of water deficit on winegrape yield and quality.* Davis, Calif.: Department of Land, Air and Water Resources, University of California.

Proffitt, T., R. Bramley, D. Lamb, and E. Winter. 2006. *Precision viticulture: A new era in vineyard management and wine production.* Adelaide: Winetitles.

Roberts, D., and A. Cass 2007. Influence of available water and salinity on rootstock selection. *Practical Winery and Vineyard* 29 (2), 34–42.

Robinson, J. B. 2005. Critical plant tissue values and application of nutritional standards for practical use in vineyards. In *Soil environment and vine mineral nutrition,* ed. L. P. Christensen and D. R. Smart, 61–68. Davis, Calif.: American Society of Enology and Viticulture.

Robinson, J. B., M. T. Treeby, and R. A. Stephenson. 1997. Fruits, vines and nuts. In *Plant analysis: An interpretation manual,* 2nd ed., ed. D. J. Reuter and J. B. Robinson, 349–382. Melbourne: CSIRO Publishing.

Seguin, G. 1986. "Terroirs" and pedology of wine growing. *Experientia* 42, 861–873.

Shackel, K. 2006. Water relations of woody perennial plant species. In *Terroirs viticoles 2006,* ed. C. van Leeuwen and J. Fanet, 54–63. Bordeaux: Vigne et Vin Publications Internationales.

Sinskey, R. 2007. Robert Sinskey vineyards, Napa California. *Daily Wine News* (www. winebiz.com.au). Adelaide: Winetitles.

Smart, R. E. 2001. Where to plant and what to plant. *Australian and New Zealand Wine Industry Journal* 16 (4), 48–50.

Smart, R. E., and P. R. Dry. 1980. A climatic classification for Australian viticultural regions. *Australian Grapegrower and Winemaker* 196, 8–16.

Smart, R., and M. Robinson. 1991. *Sunlight into wine: A handbook for winegrape canopy management.* Adelaide: Winetitles.

Smith, T. 2007. Terroirs of the Barossa. *Daily Wine News* (www.winebiz.com.au). Adelaide: Winetitles.

Soil Survey Division Staff. 1993. *Soil survey manual.* U.S. Department of Agriculture handbook no. 18. Washington, D.C.: U.S. Government Printing Office.

Steiner, R. 1993. *Spiritual foundations for the renewal of agriculture: A course of lectures.* Kimberton, Penn.: Bio-Dynamic Farming and Gardening Association.

Suzuki, D. 1998. *Earth time.* St. Leonards, New South Wales: Allen and Unwin.

Thorn, R. G. 2000. Soil fungi. In *Handbook of soil science*, ed. M. E. Sumner, C-22–C-37. Boca Raton, Fla.: CRC Press.

Webb, L. B., P. H. Whetton, and E. W. R. Barlow. 2006. Potential impacts of projected greenhouse gas-induced climate change on Australian viticulture. *Australia and New Zealand Wine Industry Journal* 21 (4), 16–20.

Wetherby, K. 2000. *Soil description book.* Clare, South Australia: K. G. and C. V. Wetherby.

White, R. E. 2003. *Soils for fine wines.* New York: Oxford University Press.

White, R. E. 2006. *Principles and practice of soil science.* 4th ed. Oxford: Blackwell Publishing.

Whiting, J. R. 2003. *Selection of grapevine rootstocks and clones.* State of Victoria: Department of Primary Industries.

Whiting, J. R. 2004. Grapevine rootstocks. In *Viticulture Volume 1 Resources*, 2nd ed., ed. P. R. Dry & B. G. Coombe, 167–188. Adelaide: Winetitles.

Wilson, J. E. 1998. *Terroir: The role of geology, climate and culture in the making of French wines.* London: Mitchell Beazley.

Wine Institute of California. 2005–2008. Potential benefits of sustainable winegrowing practices (www.wineinstitute.org/initiatives/sustainablewinegrowing/benefits).

Zhang, X., R. R. Walker, R. M. Stevens, and L. D. Prior. 2002. Yield–salinity relationships of different grapevine (*Vitis vinifera* L.) scion–rootstock combinations. *Australian Journal of Grape and Wine Research* 8, 150–156.

Further Reading

Dry, N. 2007. *Grapevine rootstocks: Selection and management for South Australian vineyards.* Adelaide: Winetitles.

Gladstones, J. S. 1992. *Viticulture and the environment. A study of the effects of environment on grapegrowing and wine qualilty with emphasis on present and future areas for growing winegrapes in Australia.* Adelaide: Winetitles.

Jackson, R. E. 2000. *Wine science: Principles, practice, perception.* 2nd ed. London: Academic Press.

Johnson, H., and J. Robinson. 2007. *The world atlas of wine.* 6th ed. London: Mitchell Beazley.

Magarey, P. A., A. M. MacGregor, M. F. Wachtel, and M. C. Kelly. 2006. *Field Guide for Diseases, Pests and Disorders of Grapes for Australia and New Zealand.* Adelaide: Winetitles.

Pomerol, C. 1989. *The wines and winelands of France. Geological journeys.* London: Robertson McCarta.

Robinson, J. (ed.) 1999. *The Oxford companion to wine.* 2nd ed. Oxford: Oxford University Press.

Winkler, A. J., J. A. Cook, W. M. Kliewer, and L. A. Lider. 1974. *General viticulture.* Berkeley, Calif.: University of California Press.

Web Sites

www.ams.usda.gov

www.australianorganic.com.au

www.biodynamics.net.au

www.bom.gov.au/silo

www.cimis.water.ca.gov

www.clw.csiro.au/publications/projects20.pdf

www.daff.gov.au/agriculture-food/food/organic-biodynamic

www.demeter.net

www.earthit.com

www.ecocert.com

www.environment.gov.au/about/councils/ephc/index.html

www.fullstop.com.au

www.geonics.com

www.ifoam.org

www.ipw.co.za

www.lodiwine.com

www.nasaa.com.au

www.natrakelp.com.au

www.nzwine.com

www.ofa.org.au

www.redwhiteandgreen.com.au

www.samdbnrm.sa.gov.au/Portals/7/AWMN/awsview.php

www.smartvit.com.au

www.virtual-museum.soils.wisc.edu/displays.html

www.vitiplus.ch/home-content.asp?siteid=vitiswiss

www.wfa.org.au

www.wineinstitute.org

Index